Chemometrics and Species Identification

Chemometrics — General Introduction and Historical Development

B. G. M. Vandeginste

Chemometrics — Sampling Strategies

G. Kateman

Signal and Data Analysis in Chromatography

H. C. Smit and E. J. van den Heuvel

Chemometrics in Food Chemistry

M. Forina, S. Lanteri and C. Armanino

Species Identification for Trace Inorganic Elements in Biological Materials

Ph. H. E. Gardiner

AKADEMIE-VERLAG BERLIN

Topics in Current Chemistry · Vol. 141

Chemometrics and Species Identification

Chemometrics and Species Identification

With 84 Figures and 19 Tables

AKADEMIE-VERLAG BERLIN
1987

Die Originalausgabe erscheint im Springer-Verlag
Berlin · Heidelberg · New York · London · Paris · Tokyo
in der Schriftenreihe „Topics in Current Chemistry",
Volume 141

Vertriebsrechte für die sozialistischen Länder:
Akademie-Verlag Berlin

Vertriebsrechte für alle Staaten mit Ausnahme der sozialistischen Länder:
Springer-Verlag Berlin · Heidelberg · New York · London · Paris · Tokyo

ISBN 3-540-17308-0 Springer-Verlag Berlin · Heidelberg · New York
ISBN 0-387-17308-0 Springer-Verlag New York · Heidelberg · Berlin

ISBN 3-05-500371-3 Akademie-Verlag Berlin

Erschienen im Akademie-Verlag Berlin, DDR-1086 Berlin, Leipziger Straße 3—4
Lizenznummer: 202 · 100/511/87
Printed in the German Democratic Republic
Gesamtherstellung: VEB Druckerei „Thomas Müntzer", 5820 Bad Langensalza
LSV 1205
Bestellnummer: 763 736 7 (3072/141)
10800

Editorial Board

Table of Contents

Chemometrics — General Introduction and Historical Development

Bernard G. M. Vandeginste

Laboratory for Analytical Chemistry, Faculty of Sciences,
Catholic University of Nijmegen, Toernooiveld
6525 ED Nijmegen, The Netherlands

Table of Contents

Analytical chemists deal with a wide range of decision making problems such as: the selection and design of an analytical method, and processing and interpretation of the measured data. For this purpose, formal strategies and algorithms based on statistical and mathematical techniques are needed. The development of such strategies is the area of research of Chemometrics, a discipline of analytical chemistry. In this paper the role of chemometrics in the analytical process is discussed and a historical survey is given of the development of Chemometrics during the past two decades. A selection of standard Chemometric tools available to the analytical chemist is discussed in some more detail: multivariate optimization, data processing and calibration. The paper is closed with a few remarks on future directions of Chemometrics.

Bernard G. M. Vandeginste

1 The Role of Chemometrics in the Analytical Process

1.1 Economical Aspects of Analytical Information

In its narrow sense, chemical analysis is an activity of obtaining information on the identity or on the quantitative composition of a sample. By chemical analysis an analytical result is produced, which may be one or more numbers, or one or more compound names. Why do analysts, or in general analytical laboratories, produce these numbers and names? This question has been addressed by several analytical chemists [1, 2, 3].

The proposed answers vary from "because everyone does" to "because we think that the analytical results contain relevant information for the customer who asked for the analysis". Another often mentioned reason is simply "because the customer asked for it". As Massart [1] pointed out, it is to be hoped that your own answer is not this last one, but is instead "because we think the value of the information present in the analytical result is more worth than the cost of obtaining it". This means that analytical information has an economical value. This fact confronts us with three problems, namely: how can we quantify the amount of information, or the quality of information present in the analytical data? What are the cost of chemical analysis? How to quantify the economical value of analytical information?

Intuitively we can feel that the economical value of the analytical result is related to its quality. The quality of an analytical result depends upon two factors: first of all we should know how confident we are about the produced result. In fact, an analytical result without an explicit or implicit (by the number of significant figures) indication of its precision has no quality at all. Second, the quality of the analytical result depends on how well the sample represents the system of its origin. The sample may be contaminated or may be modified because of inappropriate storage and aging. In other instances, when the sample is taken from a chemical reactor in which a chemical reaction is occurring, the constitution of the reactor content is usually time varying. Because of inevitable time delays in the analytical laboratory, the constitution of the sample will not anymore represent the actual constitution in the reactor at the moment when the analytical result is available. Therefore, both the precision of the analytical method and the analysis time are important indicators for the quality of an analytical result [4].

This requirement of being able to attach a quality label to our analytical results, made that statistics and the statistical treatment of our data have become of a tremendous importance to us. This is reflected by the fact that in 1972 ANALYTICAL CHEMISTRY started with the publication of a section on "Statistical and Mathematical Methods in Analytical Chemistry" [5, 6] in its bi-annual reviews. Although we feel us quite confident on how to express our uncertainty (or certainty) in the produced numbers, we are less sure on how to quantify our uncertainty in produced compound names or qualitative results.

The economical value of the analytical result depends upon the amount of information the customer actually receives, and upon whether the customer indeed uses that information. The amount of received information can be defined as the difference between the initial uncertainty (H_0) of the customer before receiving the analytical result(s) and the remainder uncertainty (H_1) after having received the result(s). The net yield of information is thus: $\Delta H = H_0 - H_1$. If we apply this definition to the case of process control, then H_0 is related to the variance of the uncontrolled process and H_1 is related to the variance of the controlled process. When considering the cost-effectiveness of analytical effort, we should therefore, weight the cost (C) of producing analytical information (H_1), against the return or profit (P), earned by applying the net amount of received information (ΔH) by the customer. Decision making in the analytical laboratory is, therefore, focussed on maximizing the net profit (P—C). This obliges the manager of the analytical laboratory to keep evaluating the analytical methods and equipment in use in the laboratory, in relation to the changing demands for information by his customers and to the new technologies introduced on the market place. Todays equipment is of an increasing sophistication, with capabilities to determine more analytes in a shorter time and with better precision and contains software to treat the complex data structures it generates.

Two examples are given below, which demonstrate the economical principles mentioned above.

The first example is from Leemans [4)] and applies to process control. When monitoring a process, the uncertainty about the actual state of an uncontrolled process is the variance of the parameter of interest: s_0^2. From information theory [7)] it follows that the initial information (I_0) available on the process is inversely proportional to the uncertainty, namely: $I_0 = \log_2 1/s_0^2$ ($\log_2$ is the logarithm on a base 2).

Equally, the uncertainty about the process value after analysis is the variance of

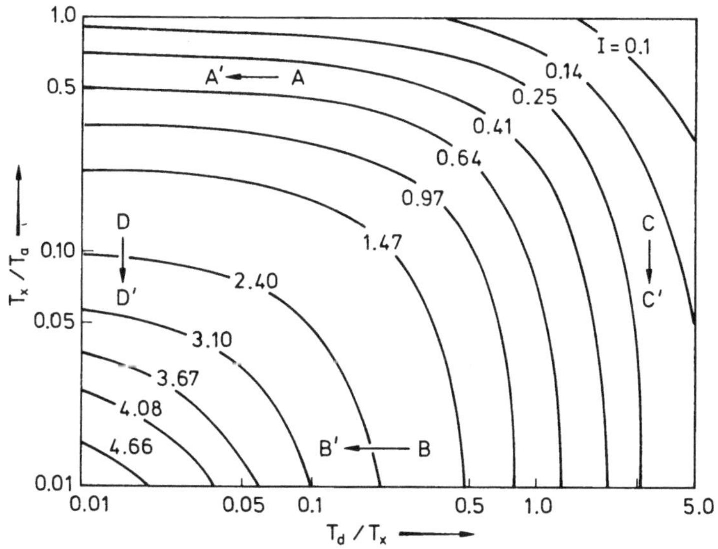

Fig. 1. The net yield of information (I), in bits, obtained by analysis with an analytical method with an analysis time, Td, and analysis frequency, 1/Ta, for the control of a process with a time constant, Tx. From G. Kateman and F. W. Pijpers, "Quality control in analytical chemistry" p. 87 (1981). Copyright © 1981, John Wiley & Sons Inc. New York. Adapted and reproduced by permission of Wiley & Sons, Inc., New York

the controlled process: s_1^2. The information after analysis is therefore, $I_1 = \log_2 1/s_1^2$. The net yield of information $\Delta I = I_1 - I_0 = \log_2 s_0^2/s_1^2$.

Leemans [4] and Müskens [8] derived a relationship between the net yield of information and the quality of the analytical procedure, expressed in terms of analysis time (T_d), precision (σ_a) and the sampling frequency ($1/T_a$). Their results are graphically displayed in Fig. 1. The conclusion on one hand is predictable but is on the other hand very striking. Equal investments in the laboratory may have different effects on the net yield of information, depending on the particular situation, e.g. an increase of the workload by a factor of 2 may have a very minor effect (point C → C′ in Fig. 1), or may have a very pronounced effect (point D → D′ in Fig. 1). The diagram also shows that a replacement of a method by a twice as fast one (e.g. as a result of optimization) may have no effect at all (point A → A′ in Fig. 1) or may have a significant effect (point B → B′ in Fig. 1). This proves that equal marginal cost may yield different marginal returns.

The second example is from Massart [1], who derived a relationship between the quality of an analytical result and its utility for medical diagnosis. As an indicator for the utility of a clinical test, one can use its specificity. This is the percentage of "normal" patients which are recognized as such. The less analytical errors are made, the better the specificity will be, which is shown in table 1. This table demonstrates

Table 1. Percentage (u) of "normal" patients recognized as such and amount of produced analytical information (I). Data from Acland and Lipton [9] and adapted by Massart [1]. Reprinted by permission of Elsevier Science Publishers, Amsterdam

S_A/S_N	u	$I = \log_2 (S_N/S_A)$
0.1	99	3.32
0.2	99	2.32
0.3	98	1.73
0.4	97	1.32
0.5	95	1.00
0.6	94	0.73
0.8	90	0.32
1.0	86	0

that the law of marginal utility applies in analytical chemistry. For the same amount of extra information, the obtained marginal utility decreases. Both examples demonstrate that although decision making in the analytical laboratory is very complex, it could be made easier when formal knowledge is available on basic relationships between the amount of generated information and the characteristics of the analytical method. In most of the cases, these relationships are expressed in mathematical or statistical formulas or models. It is, therefore, necessary to try to formalize the different parts of the analytical process.

1.2 Stages in the Analytical Process

Despite the large amount of different analytical procedures, a regularily returning pattern of activities can be discovered in chemical analysis. This pattern of activities can be considered to define the analytical process. In the previous section it was explained that the analytical system, which consists of a sample input and result output (Fig. 2) represents only a part of the total analytical process. The sample is the result of several actions after the receipt of the customers' chemical problem. The obtained analytical result, however, still needs to be converted into information. Therefore, the analytical process is better described as a cycle, which begins with the formulation of the problem that needs a solution by chemical analysis and is finished after

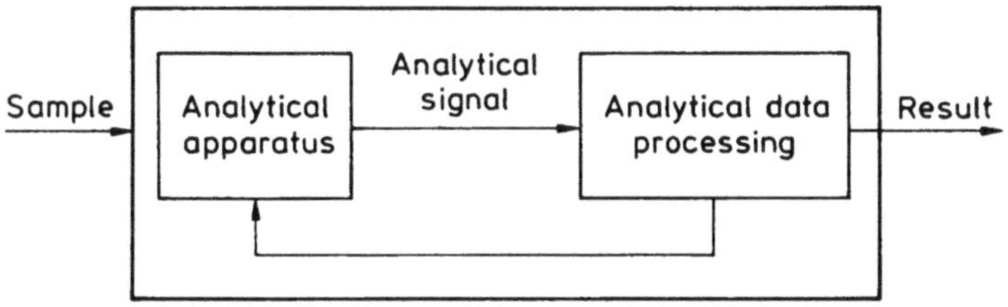

Fig. 2. The analytical method. From K. Eckschlager, V. Stepanek, Anal. Chem. *54*, 1115A (1982). Reproduced by permission of the American Chemical Society, Washington DC.

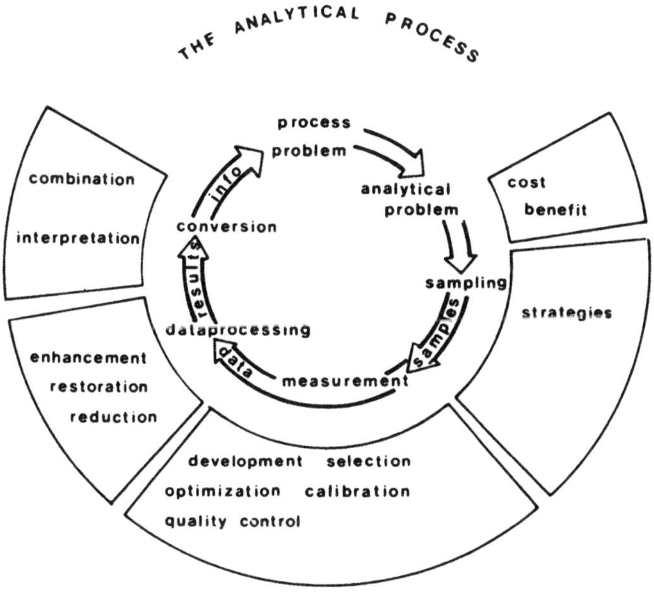

Fig. 3. The analytical process. From B. G. M. Vandeginste, Anal. Chim. Acta *150*, 201 (1983). Reproduced by permission of Elsevier Science Publishers, Amsterdam.

the analytical information has been actually supplied to solve that problem (Fig. 3). By a number of consecutive actions (Fig. 3) the initially formulated problem is translated into an analytical problem, (a) sample(s), rough data, refined data, analytical results, information and finally knowledge. The activities necessary for these conversions are: method selection, sampling, sample preparation and measurement, data processing and finally results processing. These steps define the principal stages of the analytical process. Associated with these steps are the decisions we have to make in order to obtain a maximal net profit. Without aiming to be exhaustive, a number of these decisions are listed below:

— method selection:
  make a list of candidate methods, evaluate cost, precision, analysis time, expected workload in view of required sampling scheme.
— sampling:
  derive the best sampling strategy.
— measurement:
  tune the initial measurement conditions for optimal performance (e.g. resolution in chromatography).
  select a proper calibration method: univariate or multivariate.
  design a system for quality control
— data processing:
  enhance the data if necessary; select the proper filter or smoothing technique
  restore the data if necessary; select the proper method for deconvolution.
  reduce the data to concentrations; select the proper univariate or multivariate method result processing:
  combine, classify and interpret the results — select the proper multivariate method.

Because the analytical process is a cycle or chain, each link or operation defines the ultimate quality of the analytical information. The effect of a poor sampling strategy will be very difficult to be compensated by a very good calibration method and vice versa. It is, therefore, the uneasy task of the analytical chemist to make the right or best decision at every stage of the analytical process. A large part of the decision process was believed being impossible to be formalized. Many have put up with the apparent fact that a successful analyst has an inexplicable sense of the right decision. This would reduce analytical chemistry to an art, which is not. It is likely that the above mentioned decisions cannot be made without the support of applied mathematics and statistics. Our possibilities to apply these techniques depend strongly on the availability of modern computer technology and on the imagination of the analytical chemist to follow closely on heels the advances in computer science, mathematics and statistics. The necessity to apply these techniques becomes the more and more urgent when analytical equipment produces the more complex data. A typical example is a new class of analytical methods, which consists of two linked methods such as gas chromatography-mass spectrometry.

In the present time with almost unlimited computer facilities in the analytical laboratory, analytical chemists should be able to obtain substantial benefits from the application of time series, information theory, multivariate statistics, a.o. factor analysis and pattern recognition, operations research, numerical analysis, linear algebra, computer science, artificial intelligence, etc. This is in fact what chemometricians have been doing for the past decades.

1.3 Chemometrics in the Analytical Process

Chemometrics is not a mathematical discipline and should not be confounded with one or more of the disciplines from mathematics. It is, however, a chemical discipline, as is schematically shown in Fig. 4. The inner circle represents the chemical analysis. The decisions mentioned in previous sections are supported by chemometric tools. Chemical analysis together with the chemometric tools belong to the area of analytical chemistry, which is schematically represented by the outer circle. The mathematical techniques surrounding this outer circle are auxilary techniques to the analytical chemist. In this picture, Chemometrics is the interface between chemistry and mathematics. As Kowalski [10] clearly stated "Chemometric tools are vehicles that can aid chemists to move more efficiently on the path from measurements to information to knowledge". This brings us to the formal definition of chemometrics [11] "Chemometrics is the chemical discipline that uses mathematical and statistical methods (a) to design or select optimal mesurement procedures and experiments and (b) to provide maximum chemical information by analyzing chemical data. In the field of Analytical Chemistry, Chemometrics is the chemical discipline that uses mathematical and statistical methods for the obtention in the optimal way of relevant information on material systems".

Key words in the definition are "optimal" and "material systems". These express the fact that chemical analysis is related to a problem and not to "a sample" and that economical aspects of chemical analysis prevail. The result of chemometric research is chemometric software, which enables a large scale implementation and application of chemometric tools in practical chemical analysis.

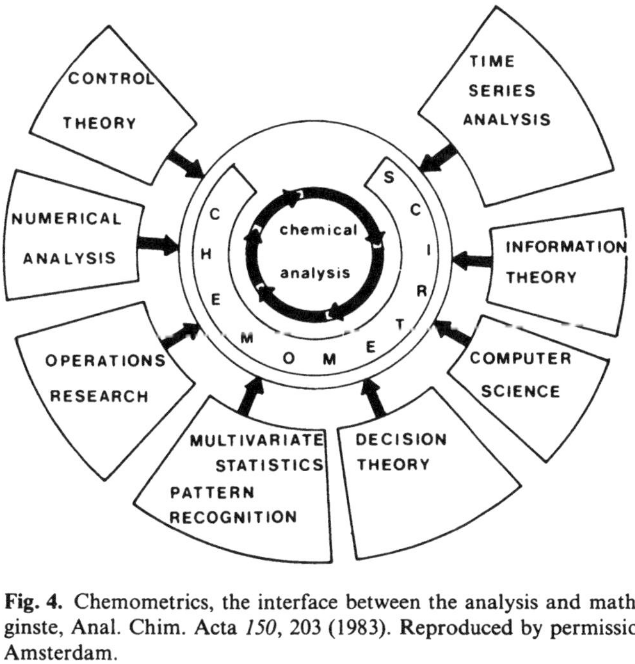

Fig. 4. Chemometrics, the interface between the analysis and mathematics. From B. G. M. Vandeginste, Anal. Chim. Acta *150*, 203 (1983). Reproduced by permission of Elsevier Science Publishers, Amsterdam.

2 Chemometrics in a Historical Context

It happens that this paper is published a year after the 10-th anniversary of the Chemometric Society, which was founded by Wold en Kowalski, who also coined at the time the word Chemometrics as a title for a new subdivision of analytical chemistry using the methods described in previous sections.

The current interest of analytical chemists may be read from the Chemometric Society Newsletter distribution list [12] (table 2).

Table 2. Chemometrics Society newsletter distribution list 1985

Country	Number	Country	Number
Australia	5	Indonesia	1
Austria	13	Iraq	1
Belgium	15	Italy	49
Brazil	3	Japan	8
Canada	12	Kenya	1
China	1	Netherlands	59
Czechoslovakia	7	Norway	29
Denmark	8	Portugal	5
Finland	5	Poland	6
France	17	Romania	4
Germany (East)	2	South Africa	1
Germany (West)	28	Spain	15
Great Britain	50	Sweden	9
Greece	2	Switzerland	4
Hungary	7	Turkey	4
India	2	U.S.S.R.	1
Yugoslavia	5	United States	262
Iceland	1		

Apparent centers with major interest in Chemometrics are the USA and the Netherlands in Europe, with respectively 262 and 59 members. Therefore, it is interesting to report on a survey made by Tuinstra et al. [13] of 300 Dutch analytical laboratories on their knowledge of, and familiarity with modern Chemometric optimization strategies. These optimization strategies require a relatively low level of abstraction and mathematics and can be expected to be fairly well known: 36% have heard or read about these strategies. Only 5% know and use optimization techniques and another 6% of the respondents have seriously considered using optimization techniques but finally decided against it. There was no difference observed between the private sector, clinical laboratories and governmental laboratories. In every case but one there was at least one university graduate present. The investigators of this survey found these figures disappointingly low. One should, however, take into account the facts that there is no real textbook available on Chemometrics and there is a considerable time lag between research and education. Another point which hindered a large scale application is the availability of certified and thoroughly tested Chemometric software at the time of this survey.

Two major developments in the past decade increased the impact of Chemometrics: the development of computers and micro-electronics and the advancement of analytical instrumentation.

In a lecture given to the Analytical Division of the Royal Chemical Society, Betteridge [14] summarized the impact of these developments on analytical chemistry as providing solutions to barriers to analytical information, giving rise to new problems (table 3). I will follow his lines to discuss the evolution of Chemometrics in its historical context.

Table 3. The generation of analytical information. From J. Betteridge, lecture presented at the Symposium "The integrated approach to laboratory automation", RCS, Analytical Division, Dorset, October 1985

	Barrier to analytical information	Solution	New problem
pre 1960	Data generation (burette)	Electronic control	Miles of chart paper
1960's	Data acquisition	Digitisers	Masses of data and results
1970's	Data/result reduction	Mini/microcomputers	Masses of information
1980's	Information management	Workstations, LIMS	Complexity of decisions
1990's	Intelligence	AI, expert systems	Fundamentals of analytical chemistry

The pre-sixties was a period just before the second phase of the electronics revolution that took place in 1960. During and before the fifties, most of our analytical equipment had to be controlled manually, making the data collection slow and laborous. The measurement of a spectrum with a single beam spectrometer, for instance, had to be carried out point by point, with in between a manual adjustment of the monochromator and baseline. The principal barrier to the production of analytical information was, therefore, the data generation. It increased the desire for having recording instruments. The first spectrometers with automatic wavelength scanning and baseline correction became widespread available in the fifties. As a consequence data generation became relatively easy and fast, causing, however, the production of miles of chart paper. The lack of the possibility to transform recorder traces into useful analytical information became the next barrier to the production of analytical information.

In the sixties semiconductor devices were introduced, changing the design of analytical instruments and dropping the price of computers by a factor 10 by the late sixties. The dedicated computer (by now called minicomputer) appeared in the bigger analytical research laboratories. Although access to such a computer was not very convenient and interfacing was not standardized and painful, the analytical laboratory could generate masses of data and results, such as digitized mass spectra, infrared spectra etc. The introduction of these dedicated (interfaced) computers was going to change the face of chemical analysis, if adapted to our needs. Before a computer can do anything with signals, they need to be converted from an analogue into a digital form. Digital data processing required the development of discrete, numerical algorithms and their translation into a computer readable code, called software. It is, therefore, not surprising that at that time much attention was given on the treatment of signals. In 1968 the Fast Fourier Transform algorithm was published [15] for the calculation of a discrete Fourier Transform. Its counterpart in the time domain, digital smoothing, was developed by Savitzky and Golay [16] and published in 1964. The application of advanced algorithms on analytical signals became easier with the publication of complete program listings. For the processing of UV-Vis and IR-spectrometry data, for example R. N. Jones [17,18,19], published quite a complete and useful package, which contained programs for curve fitting, the calculation of derivative spectra, spectrum subtraction etc. The underlying algorithms were selected after a careful investigation of performance.

The seventies are marked by tumultuous developments. Computers evolved from an expensive tool for the few to a cheap, everywhere present tool for everybody. In 1978 the first complete microcomputersystem was introduced, first with relatively modest capabilities but later (80-ies) getting more calculating power than the obsolete mainframe from the sixties. Of course, the advances in digital electronics would also influence analytical instruments. Analogue meters were first replaced by digital BCD-displays. Later switches and buttons were replaced by a keyboard. At the 1975 Pittsburgh Conference on Analytical Chemistry and Applied Spectroscopy, the first analytical instruments appeared which where operated under complete control of a microprocessor. These included an electrochemical system, a number of X-ray fluorescence systems, a mass spectrometer, and a programmable graphite furnace atomic absorption instrument. Besides the process of further sophistication of existing measuring principles, new types of devices for chemical analysis were introduced. For example, a chromatograph coupled to a mass spectrometer. This new type of equipment, by now called hyphenated method, generates no longer one spectrum or one chromatogram, but is capable to measure several spectra at short time intervals. The data form a data matrix. When operating such an instrument during 15 minutes, with one spectrum per second, digitized over 200 mass units, this data matrix contains 180.000 datapoints! If one data point occupies 4 byte, 720 Kb information has been collected. The impact of these hyphenated systems in analytical chemistry can be read from Table 4, which shows the state of the art by the beginning of the 80-ties 52 different hyphenated methods are available and 16 new methods are expected to appear in the eighties [20].

An eversince lasting development in analytical chemistry, of course, is the introduction of new techniques. Inductively coupled plasma atomic emission in the seventies, is an example. Obviously a new barrier to the production of analytical information was the problem of how to transform these masses of data into accurate and reliable information. Equally, a growing need was felt to evaluate and optimize analytical methods and procedures for obtaining a cost-effective laboratory operation. This was the ground, fertilized by spectacular developments in micro-electronics, on which a new discipline in analytical chemistry, Chemometrics, was born. Early Chemometric research (sometimes without using that name) was concentrated in the U.S.A., Sweden, Belgium and the Netherlands. The first paper mentioning the name "Chemometrics" was from Wold and was published in 1972 in the Journal of the Swedish Chemical Society. At the same time, the "Arbeitskreis Automation in der Analyse" initiated a discussion on a systems approach of analytical chemistry, in West-Germany. Chemo-

Table 4. The state of the art of the Hy-phen-ated methods. From T. Hirschfeld, Anal. Chem. *52*, 299A (1980). Reprinted by permission of the American Chemical Society, Washington DC

▒ Requires further invention
● Feasible in the state-of-the-art
□ Presently successful

	Gas chromatography	Liquid chromatography	Thin layer chromatography	Infrared	Mass spectroscopy	Ultraviolet (visible)	Atomic absorption	Optical emission spectroscopy	Fruorescence	Scattering	Raman	Nuclear magnetic resonance	Microwaves	Electrophoresis
Gas chromatography	●	▒	▒	□	□	●	▒	●	▒	▒	●	●	●	▒
Liquid chromatography	●	▒	▒	●	□	□	●	●	□	□	●	●	●	▒
Thin layer chromatography	●	▒	□	●	●	●	▒	▒	●	▒	●	▒	●	●
Infrared	▒	▒	▒	▒	●	●	▒	▒	●	▒	▒	●	●	▒
Mass spectroscopy	▒	▒	▒	●	□	●	▒	●	▒	▒	▒	●	●	▒
Ultraviolet (visible)	▒	▒	▒	●	●	▒	▒	▒	●	●	▒	●	●	▒
Atomic absorption	▒	▒	▒	▒	▒	▒	▒	●	▒	▒	▒	▒	▒	▒
Optical emission spectroscopy	▒	▒	▒	▒	●	▒	●	▒	▒	▒	▒	▒	▒	▒
Fluorescence	▒	▒	▒	▒	▒	●	▒	▒	□	□	□	▒	▒	▒
Scattering	▒	▒	▒	▒	▒	●	▒	▒	□	▒	□	▒	▒	▒
Raman	▒	▒	▒	▒	▒	▒	▒	▒	●	●	▒	▒	▒	▒
Nuclear magnetic resonance	▒	▒	▒	●	●	●	▒	▒	▒	▒	▒	□	▒	▒
Microwaves	▒	▒	▒	▒	●	▒	▒	▒	▒	▒	▒	▒	▒	▒
Electrophoresis	▒	▒	●	▒	▒	●	▒	▒	●	□	▒	▒	▒	□

metrics research in the U.S.A. and Sweden was from the early beginning focussed on the application of pattern recognition and related techniques as factor analysis and on the optimization of analytical procedures.

In 1969 Jurs, Kowalski and Isenhour published a first series of papers in ANALYTICAL CHEMISTRY, reporting the results of applying the linear learning machine to low resolution mass spectral data [21,22,23]. The goal of these studies was to extract molecular structural information directly from spectral data. The first application of the SIMPLEX method for a sequential optimization, which was developed in the early 1960's [24], dates from 1969 [25] and was picked up soon by several others. Deming [26] investigated in detail the applicability of the SIMPLEX method for the optimization of a wide variety of analytical procedures. The Dutch-German-Belgian research was more focussed on a systems approach of the information production in the analytical laboratory.

Gottschalk [27] and Malissa [28] attempted repeatedly to systemize chemical analysis. They introduced a symbolic iconic language, the SSSAC-system (Symbolic Synoptic System for Analytical Chemistry), but failed to get their system introduced. A large part of their problem was that the SSSAC-system could not be translated into a computer readable code.

Table 5. A computer readable code of the symbolic representation of the complexometric titration of ferric iron in Renn Slag, given in Fig. 5. From H. Malissa and G. Jellinek, Z. Anal. Chem. *247*, 4 (1969). Reprinted by permission of Springer Verlag, Berlin

Step	Command	Comment
1	samp S 1	
2	add L 1	
3	solv (1200)	1200 = 20 hrs
4	filt	
5	wash L 2	
6	add L 3	
7	heat (100; 60)	100 °C during 1 hr
8	dilut L 4 (150)	dilute to 150 ml
9	add S 2	
10	add L 5	
11	if (PH.LT.2.0) GO TO 10	
12	titr L 6 (70)	titrate on 70 °C
13	END	

L 1, L 2, L 3, L 4, L 5, and L 6 are solutions; S 1 and S 2 are solids

Nevertheless, their concepts at that time were very close to the way robots are programmed (table 5) and automatic analyzers are controlled by now. As a comparison, a FORTH program is shown (table 6) for the automatic control of a UV-Vis spectrometer [29]. A revival of their approach is not unthinkable, having now symbolic computer languages, such as PROLOG and LISP, available for the manipulations of knowledge and FORTH for instrument control. Figure 5 shows the symbolic representation of a complexometric titration of ferric iron in Renn Slag [28] with potentiometric indication. The corresponding computerprogram, proposed by Malissa [28] is given in table 5, with an apparent resemblance with a program written in FORTH (table 6).

Table 6. A program written in FORTH, to drive a UV-VIS spectrometer

Step	Command
1	%T zero
2	SET.MONO
3	START.STOP
4	SCAN.RATE.10
5	360 760 STEP.WAVE
6	END

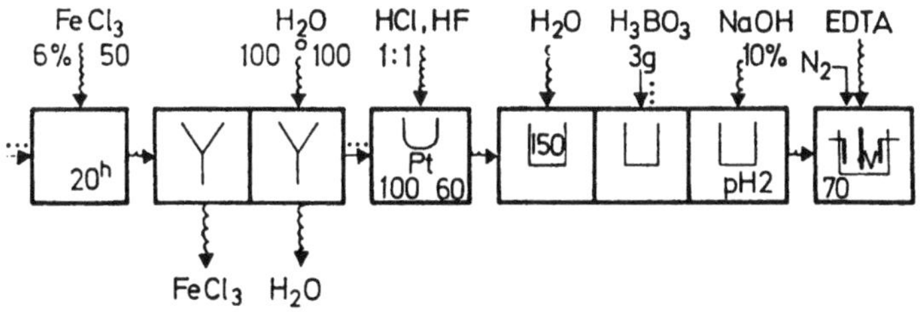

Fig. 5. A symbolic representation according to the Symbolic Synoptic System for Analytical Chemistry (SSSAC) of the complexometric titration of ferric ion in Renn Slag. From H. Malissa and G. Jellinek, Z. Anal. Chem. *247*, 3 (1969). Reproduced by permission of Springer Verlag, Berlin.

One of the aims of the systems approach of the information production process in the analytical laboratory was to quantify the information content of an analytical result. As a result, there was a renewed interest in applying information theory to analytical chemistry, with centers of research in the laboratories of Eckschlager [30)] in Czechoslovakia and Dijkstra [31)] in the Netherlands. Although information theory has not really led to new concepts in quantitative analysis, it has considerably aided to predict the effect of combining analytical procedures [32)]. Because storage capacity and speed of computers were limited at that time, compared to todays standards, it was important to design optimal coding rules for spectral data (mass and infrared spectra) and to design optimized retrieval procedures. This could be realized by the application of information theory [33)]. Having the opinion that the optimization of the information production should include the whole laboratory and not be limited to the separate analytical methods, Massart [34)] and Vandeginste [35)] investigated the applicability of Operations Research to formalize the decision making process in the analytical laboratory. Ackoff and Sasieni [36)] defined Operations Research as "the application of a scientific method by interdisciplinary teams to problems involving the control of organized (man-machine) systems so as to provide solutions which best serve the purposes of the organization as a whole". Massart et al. applied methods such as graph theory [37)], branch and bound methods [38)] and dynamic programming [39)] to optimize analytical procedures. Vandeginste [40)] applied discrete event simulation to predict the effects of the various options for the organization of sample streams, equipment and personel on delays in the laboratory.

In the mid-seventies multivariate methods, such as pattern recognition and factor analysis were gaining an increasing importance in analytical chemistry. The necessity was felt for a rapid exchange of experiences, results and computer programs among Chemometricians. This gave the sign for the necessity of a Chemometrics Society, which was started by Wold and Kowalski in 1974.

By the end of the seventies, Shoenfeld and DeVoe [6)] provided the editor of ANALYTICAL CHEMISTRY with the new title "Chemometrics" for the bi-annual review "Statistical and Mathematical Methods in Analytical Chemistry". This was a formal recognition that a new subdiscipline in Analytical Chemistry was born, which was emphasized by the special attention on Chemometrics at a symposium on the occasion of the celebration of ANALYTICAL Chemistry's 50-th anniversary [41)].

In 1977 the American Chemical Society organized its first symposium on Chemometrics: Theory and Application [42]. One year later a textbook on Chemometrics was published [34] entitled "Evaluation and optimization of laboratory methods and analytical procedures". At that time, Chemometricians became fully aware of the fact that a condition for moving the Chemometric techniques from the specialized research groups to the general research and finally to the routine lab, is an easy access to software. Therefore, we cannot overestimate the impact of the software package for pattern recognition "ARTHUR" developed by Kowalski's group [43] at the University of Washington in Seattle. For many freshmen Chemometricians "ARTHUR" [44] was the first exposure to Chemometric methods such as: principal components analysis, linear learning machine, hierarchical clustering, K-nearest neighbour classification and many others.

Just before the turning point of the sevienties, easy-accessible, low cost microcomputers appeared on the market, announcing the invasion of the micro's in the analytical laboratory. The fact that Chemometrics could now be applied routinely triggered a big interest in the subject and a big demand of Chemometric education [45,46]. At the beginning of the eighties the microcomputers were still modest: 8 bit word length, 64 KBytes RAM-memory. Nowadays personal supermicrocomputers are equipped

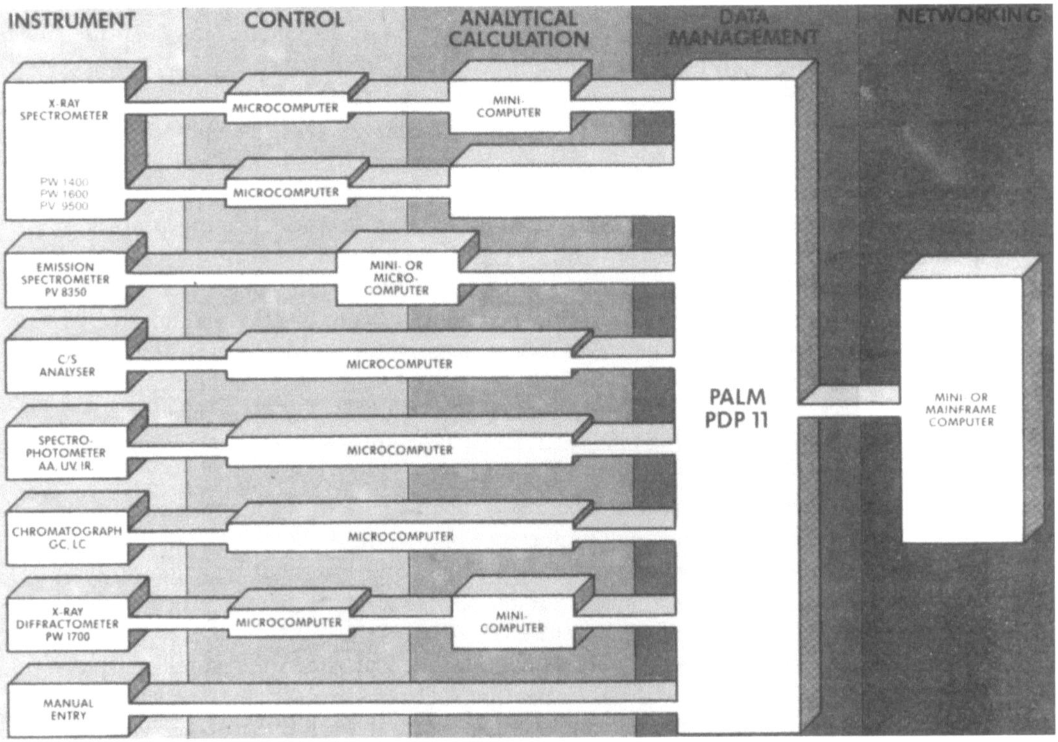

Fig. 6. A computer integrated laboratory. Reproduced by permission of Philips Netherlands

with 32 bit processors, several megabytes of RAM-memory, tens to hundreds of megabytes of disk storage capacity and several terminal ports. On-line data acquisition and real time data processing became more rule than exception. As a consequence the analytical laboratory could produce an enormous amount of information. Each analytical instrument stores the rough data, intermediate and final analytical results on discette or hard disc, with the danger of easily losing the capability to manage the data-, result- and information-streams. Although we were capable to produce good analytical information, we got the problem of becoming incapable to provide that information to the customer, because it is spread all over the place in bits and pieces. It confronted us with a new barrier to analytical information: the management of information. This reflects todays situation. An aid for controlling information flows and sample flows is the so-called laboratory-information management system (LIMS), or local area computer network (LAN) as discussed by Dessy [47,48]. A preferred organization is based on a hierarchical structure centered on a large computer connected to progressively smaller processors with the specifics tailored to the demands of the correspondent instruments and users. The central computer is dedicated to larger data processing and the real-time needs of a laboratory instrument is shouldered by microcomputers.

A typical set-up is shown in Fig. 6. The analytical instruments are coupled to a mainframe computer over a decreasing hierarchy of microcomputers and instrument processors. Figure 7 shows schematically the information flows and storage of a

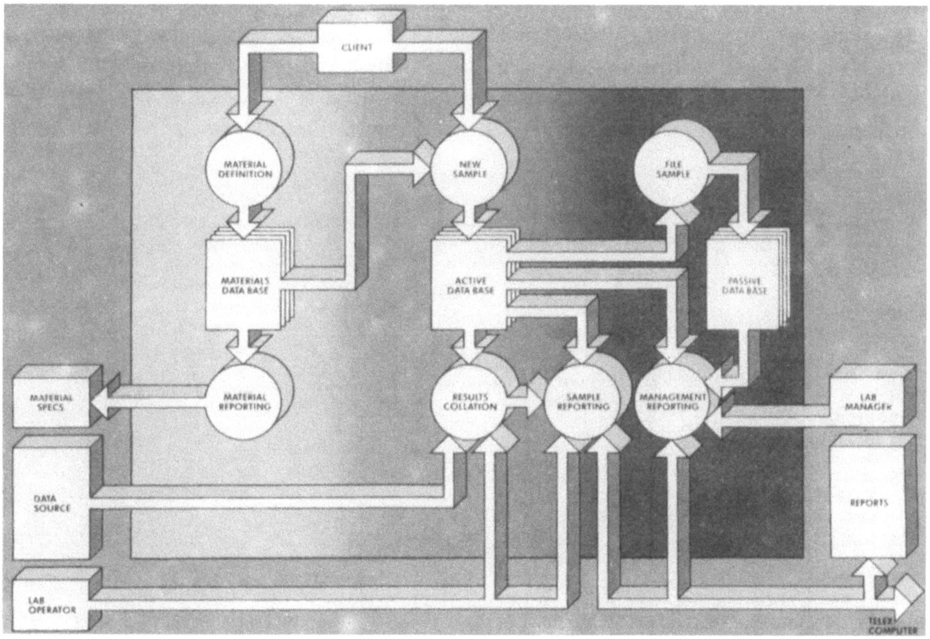

Fig. 7. The Philips Automated Laboratory Management (PALM). Reproduced by permission of Philips Netherlands

laboratory information management system, which contains modules for day-to-day sample management (the active database) and modules for the retrieval of management information on all aspects of laboratory and plant performance (the passive database).

So far, the automated analytical instrument remained a dumb piece of equipment. It repeats a sequence of actions without feedback of its readings to its functioning. The instrument cannot decide that it is out of specification and, therefore, should be retuned or recalibrated. On the other hand, if the instrument has provisions for self-diagnosis, it has usually not the necessary intelligence for a corrective action. The tuning of the instrumental conditions by an optimization procedure is still lenghty and requires the operators' continuous attention. E.g. when optimizing the separation of a mixture by HPLC, the operator has to interpret the separation obtained for a given constitution of the eluent, feed these results into a computer loaded with a program for optimization and set the new, hopefully better conditions suggested by the program into the intrument and wait for the next result. If this sequence is fully automated and under computer control, the instrument is self-optimizing. Self-optimization can be considered to be an essential element of the intelligent analytical instrument. Its principle is shown in Fig. 8. Other desirable capabilities of an intelligent instrument are selfcalibration and an automatic error diagnosis and correction. Research on self-optimizing instruments was mainly concentrated in Europe. The first instrument was a Flow Injection Analyzer (FIA), developed at Betteridges' [49] laboratory in Swansea. It adjusted automatically the flows of the carrier solution and reagents for maximal sensitivity (Fig. 9). A prototype of a self-optimizing furnace atomic absorption spectrometer was developed in Kateman's lab in Nijmegen [50], and a self-optimizing HPLC, based on the work of Berridge [51,52] is now commercially available. Self-calibrating systems are still in their infancy. Kateman and his co-workers [53,54] have demonstrated that self-calibration is in principle feasible by the application of a recursive filter, such as the Kalman filter. Thijssen [53,54] designed a

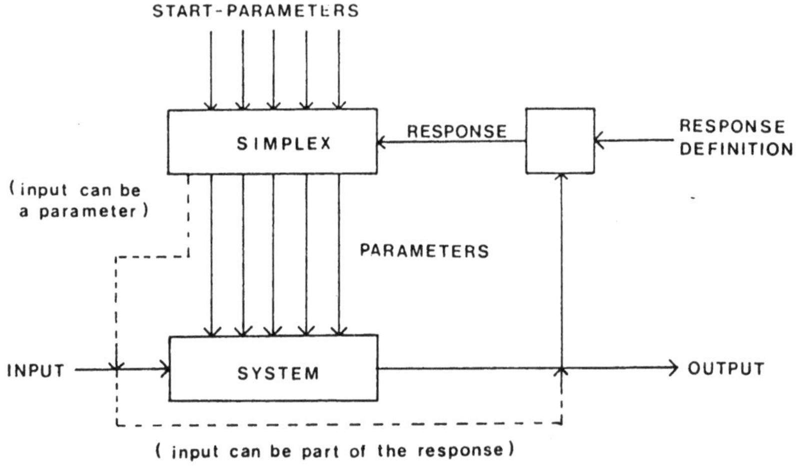

Fig. 8. Scheme of a sequential self-optimizing Analytical Instrument

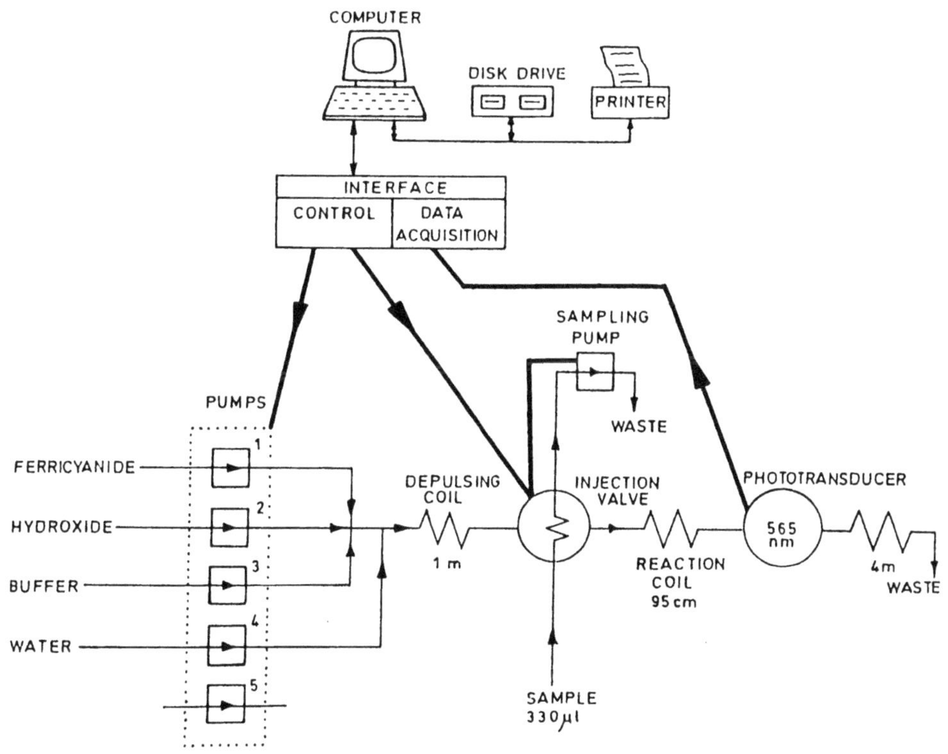

Fig. 9. A self optimizing Flow Injection Analyzer (FIA). From J. Betteridge, Lecture presented at Analysis 84, London. Reproduced by permission of the author

prototype self-calibrating FIA system for the determination of Cl^- in water. The system autonomously monitors its uncertainty about the calibration factors during the time it measures the unknown samples. On an evaluation of that precision, the instrument autonomously decides to break the measurements of the unknowns and to start the measurements of standards. The measurements of the standards are continued until the certainty about the calibration factors drops below the requested level. This cycle is schematically shown in Fig. 10. Other applications of parameter estimation by a recursive filter have been reported by Brown [55,56].

An extension of the concept of intelligent instruments is the intelligent workstation in a computer integrated laboratory. Using the word "artificial intelligence" in this context is obvious. The application of artificial computer intelligence in chemistry has not historically been considered chemometrics. However, the topic is developing rapidly in parallel with chemometrics. Traditionally, AI approaches have been restricted to problems of structure elucidation of molecules from spectral information. A well-known application of AI, also among knowledge engineers, is DENDRAL [57] for the structure elucidation of organic compounds. Another large field of applications of AI in chemistry is organic synthesis planning. Well-known systems are SECS [58] and LHASA [59]. Having these successful examples in mind and experiencing a growing need

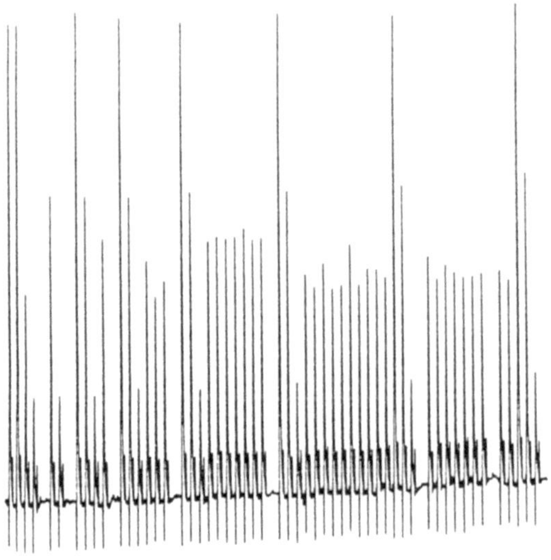

a

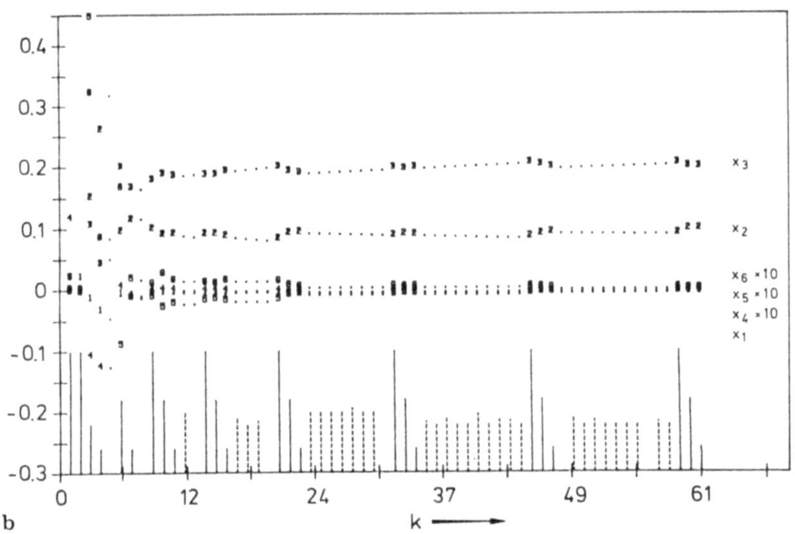

b

Fig. 10 a and b. A self calibrating Flow Injection Analyzer (FIA); **a**. Sequence of measurement of standards and samples. **b**. Schematic representation of the sequence given under (**a**). Solid lines are the readings of the standards and dashed lines are samples. x_1, ... x_3 are the estimated model parameters of the calibration function. From P. C. Thijssen, G. Kateman, H. C. Smit, Trends in Anal. Chem. *4*, 72 (1985). Reproduced by permission of Elsevier Science Publishers, Amsterdam

for converting information into knowledge to make the best decisions, a group of Chemometricians decided to explore the applicability of artificial intelligence in analytical chemistry. Not in the traditional field of structure elucidation, but for the computeraided design of analytical receipecs [60], for advice in liquid chromatography [61], for the determination of the composition of rocks from X-ray analysis [62], for sampling and trouble shooting [63] and finally for sample routing and scheduling in the analytical laboratory [64].

Although the intelligent analytical laboratory is far from reality, the new problems it will cause are already visible. Expert systems are based on knowledge and rules, which describe the analytical system (HPLC, AAS, ...). The formulation of these rules requires a profound knowledge of the system and in fact requires to return to the fundamentals of analytical chemistry. Why are analytical procedures designed as they are? Which physical and chemical laws and rules govern the analytical operation? To give an idea of the structure of a rule-base, one of the rules of a system for advice in atomic absorption spectrometry, developed by Vandeginste et al. is given in table 7.

Table 7. A production rule of an expert system for Atomic Absorption Spectrometry

```
(rule 048 curtechprep; self-ref
    (and (equalsp (best-ofp (cntxt tech-ident))
                          (quote flame))
        (biggerp (cntxt needed-dilution) (quote 500)))
    (do all (compute (cntxt needed-dilution)
            (div $ (cntxt needed-dilution)
                (quote 8)))
            (compute (cntxt burner-angle)
                    (quote 90 degrees))))
```

translation: if the best technique found so far is "flame" and the needed dilution factor found so far is greater than 500, then decrease the needed dilution-factor by a factor of 8 and turn the burner head by 90 degrees

Until now Chemometrics has been considered to be the interface between applied mathematics and analytical chemistry. I believe that by now Chemometrics is also becoming the interface between the analytical chemist and analytical chemistry. This symbiosis of artificial and natural intelligence in analytical chemistry probably can provide an optimal solution [65]. In this respect, I like to quote what Meglen [66], secretary of the Chemometric Society, wrote on the occasion of the tenth anniversary of the Society: "Chemometrics is clearly beyond infancy. Indeed at ten years old it is remarkably mature". A sign of its maturity the fact that two publishers decided to publish a journal devoted to chemometrics.

3 Examples of Chemometric Research

In this section some illustrative Chemometric results in the fields of optimization, data processing and calibration are discussed in some more detail. It should be realized, however, that these topics represent only a very small fraction of Chemometric research.

3.1 Optimization

Before formal methods for optimization were known, analytical chemists and presumably scientists in other fields, optimized systems by varying one parameter at a time. Figure 11 shows the lines with equal response for a system of which the optimization criterion depends upon two parameters x_1 and x_2. It demonstrates that this approach may fail to find the optimum. The reason for this failure is that the optimum for one parameter may depend upon the level of one or more other parameters. It is, therfore, necessary to vary all parameters — usually called factors — simultaneously. This is a multivariate approach of optimization. There are two groups of multivariate optimization methods: simultaneous and sequential methods. When using a simultaneous optimization method one carries out all experiments in parallel. The design of the experiments is factorial-type [67]. When using a sequential method, the experiments are carried out in sequence, with a feedback between the results of the experiments and the factor combinations to be selected. The sequential method which has been applied the most in analytical chemistry, is the Simplex method [24,25,68] which should not be confused with the "Simplex tableau" used in linear programming or with "Simplex mixture designs" which are a type of constrained factorial design. In a Simplex optimization, the experiments are arranged in a geometrical figure (Fig. 12). called Simplex. The coordinates of the corners of a Simplex represent the factor combinations which are selected. By dropping the corner with the worst response and by moving into the opposite direction, a new Simplex is obtained. The coordinates of the new corner represent the factor combination to be chosen next. A typical progress of a Simplex optimization of a system with two factors, is shown in Fig. 12.

The movements of a Simplex in the direction of the optimum are defined by a number of logical rules, which are applicable to systems with any number of factors. The rules and calculations for moving the Simplex can be found elsewhere [70].

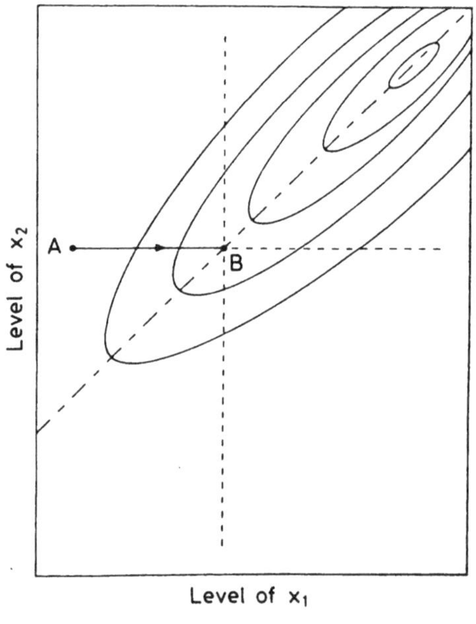

Fig. 11. The response surface of a two-factor system. The lines represent equi-response lines. Optimization by varying one factor at a time. From P. J. Golden and S. N. Deming, Laboratory Microcomputer *3*, 44 (1984). Reproduced by permission of Science & Technology Letters, England

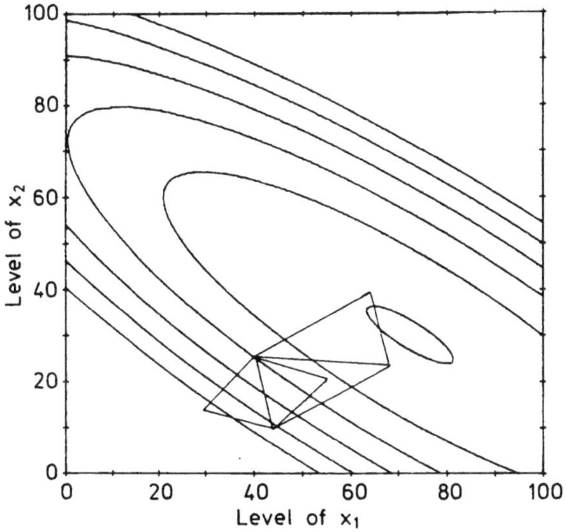

Fig. 12. The progress of the modified Simplex method for optimization. From P. J. Golden and S. N. Deming, Laboratory Microcomputer, *3*, 44 (1984). Reproduced by permission of Science & Technology Letters, England

Already in 1955, Box [71] mentioned that an evolutionary operation (EVOP) type method could be made automatic. Although the Simplex method has been critisized at many occasions, because it cannot handle situations with multiple optima or with excessive noise, its unique suitability for unattended and automatic optimization of analytical systems, explains the great effort by Chemometricians to make the method work.

The need for formal logics in optimization and the need for unattended optimization is probably the largest in chromatography, especially in liquid chromatography. Figure 13 shows a schematic representation of a chromatographic system with the controllable and uncontrollable, or fixed factors [72]. The output signal is a sequence of clock-shaped peaks, which represent the separated compounds. A first problem encountered in optimization is to decide which parameter or criterion will be optimized. In spectrometry, the criterion is more or less obvious: e.g. sensitivity. In chromato-

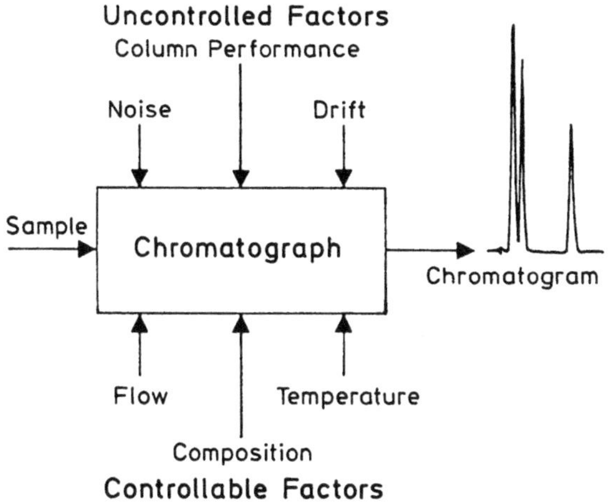

Fig. 13. A Chromatographic System. From J. Berridge "Techniques for the automated optimization of HPLC separations", 1984, J. Wiley & Sons, England, page 20. Reproduced by permission of Wiley & Sons, England

graphy, the criterion is also obvious, namely resolution. It is, however, much less obvious how to express "resolution" in one number. Since 1960 there have been numerous attempts to design quality criteria for chromatographic separation. It is generally agreed that the quality criterion should include the separation of different pairs of peaks, to produce a chromatographic Response Function (CRF), first introduced by Morgan and Deming [73]. Table 8 gives a compilation of the various chromatographic quality functions, used for a Simplex search [72].

Table 8. Chromatographic separation quality functions used for a Simplex search. From J. Berridge, "Techniques for the automated optimization of HPLC separations", p. 26–27 (1984), Wiley & Sons, England. Copyright © 1984, John Wiley & Sons, Inc. Reprinted by permission of J. Wiley & Sons Ltd. England

Chromatographic response and optimization functions	Experimental variables	Optimization method	Ref.
$CRF = \sum_{i=1}^{n} P_i$	GLC	Simplex	73)
$CRF = \frac{1}{t} \sum_{i=1}^{n} P_i$			
$COF = \sum_{i=1}^{n} a_i \ln P_i$			
$COF = \sum_{i=1}^{n} a_i \ln (R_i/R_d) + b(t_m - t_n)$	Composition of quarternary mobile fase	Simplex	76)
$CRF = \sum_{i=1}^{n} R_i + n^a - b\lvert t_m - t_n \rvert - c(t_0 - t_1)$	Composition of ternary mobile phase, temperature, flow rate, pH	Simplex	74)
$CRF = \sum_{i=1}^{n} \ln (P_i/P_0) + a(t_m - t_n)$	Gradient + parameter and flow rate		77)
$CRF = \frac{1}{t} \prod_{i} f_i/(g_i + 2n_i)$	Concentration of organic modifier and buffer, pH		78)
$CRF = \sum_{i=1}^{n} \ln (f_i/g_i) - 100(M - n)$	Concentration of organic modifier, pH		79)
$F_{obj} = \sum_{i=1}^{n} [10(1 \cdot 5 - R_i)]^2$	Composition of ternary mobile phase		80)
$F_{obj} = \sum_{i=1}^{n} 100 \cdot e^{(1 \cdot 5 - R_i)} + (t_m - t_n)^3$	Composition of ternary mobile phase		81)

P_i: peak separation for the i-th pair of peaks
a, b: arbitrary weighing factors
t_m: maximum acceptable analysis time
t_n: elution time of the last peak
R_i: resolution of the i-th pair of peaks
R_d: desired resolution for the i-th pair
P_0: the desired peak separation
n: the number of peaks detected
t_1: the elution time of the first detected peak
t_0: a specified minimum desired retention time for the first detected peak
CRF: chromatographic response function
COF: chromatographic optimization function
f/g in related to the peak separation

Table 9. Optimization experiments in HPLC using the Simplex procedure. Adapted from J. Berridge "Techniques for the automated optimization of HPLC separations", p. 150 (1984), Wiley & Sons, England. Copyright © 1984, John Wiley & Sons, Inc. Reprinted by permission of J. Wiley & Sons Ltd. England

Application[a]	Variables	Ref.
Ion-exchange separation of inorganic cations	Binary mobile phase composition	82)
R-P separation of PTH-amino acids	4 gradient parameters and flow rate	77)
R-P separation of alkaloids	Quaternary mobile phase composition and flow rate	81)
Normal-phase separation of carotenoids	Ternary mobile phase composition	80)
Fully automated examples		
Isocratic R-P separation of pyridines	Binary mobile phase composition and flow rate	74)
Gradient R-P separation of antioxidants	3 gradient parameters	74)
Isocratic R-P separation of substituted aromatics	Ternary mobile phase composition	74)
Isocratic normal-phase separation of aromatics	Ternary mobile phase composition	74)
Rapid R-P isocratic separation of sulphonamides	Ternary mobile phase composition	75)

[a] R-P = reversed-phase

The first results of optimization in chromatography were published in 1975 [73,82]. Since then a growing number of optimization experiments in HPLC using the Simplex procedure has been reported [72] (table 9). The examples are mainly reversed-phase separations, in which the composition of the ternary or binary mobile phase composition is optimized. The factors optimized are usually a selection from: flow rate, column temperature and length, the eluents constitution (e.g. organic modifier content, buffer concentration and pH), the gradient shape. Seven years after the first applications of Simplex optimization had appeared, the first fully automated optimization of HPLC separations was published by Berridge [74] in 1982. This development coincided with the introduction of fully automated optimization of Flow Injection Analysis (FIA) by Betteridge [49] and in furnace atomic absorption spectrometry by Van der Wiel [50]. Another two years later (1984) the first self-optimizing microcomputer controlled chromatograph was commercially available [51,52].

Because the logics of the Simplex method does not contain specific chromatographic knowledge, the optimization of completely unknown mixtures may fail because a local optimum is found or because the Simplex was started too far off the optimum. Berridge [72] has shown that by the inclusion of knowledge on reversed-phase chromatography in the logics, one can restrict the search area to a region of the factor space, in which the global optimum must lie. This improves considerably the speed of reaching the optimum. This principle was imbedded in a program FASTOPT [75], which calculates first the solvent strengths that will provide a reasonable reversed-phase isocratic separation. These solvent strengths are estimated from the elution times of the first and last detected peaks of a reversed-phase gradient separation using methanol and water. This algorithm was applied with success on a mixture of five sulphonamides which was believed being very difficult to optimize by sequential methods. With the example of sequential optimization by Simplex in chromatography, only a very modest part of Chemometric research on optimization has been discussed. Other strategies based on formal experimental designs and on semi-empirical relationships have been extensively investigated for the optimization of a wide variety of analytical systems.

3.2 Multivariate Data Processing

In section 2 it has been argued that the measured data structures are becoming increasingly complex which require a continuous effort to develop proper data processing methods. The complexity of the data structures is not the only factor which defines the data processing method that should be selected. Two other important factors are the available knowledge on the object, subjected to an analysis, and the expected problems during the analysis. They define how much "mathematics" should be applied to find the solution of the analytical problem at hand. This is schematically shown in Fig. 14 for the analysis of mixtures. A selection of expected problems is e.g. the occurrence of matrix effects, interferences or combination of these two. The available knowledge about the sample may vary from the number and identity of the compounds to no knowledge at all. Two other factors which dictate the complexity of the required analytical method and data processing method are the homogeneity of the object in time or position and the availability of standards or standard mixtures. When the object is homogenous, all samples will have the same constitution. If not, a certain variation in the constitution of the samples will be observed. Each combination of available knowledge, expected problems, homogeneity of the object and availability of standards, requires a proper data processing method, as is indicated in Fig. 14. For example, for homogenous objects only one specific spectrum or chromatogram is usually available. The requested analytical information has, therefore, to be derived from a single signal record. This is called a univariate analysis. Consequently, the applicable data processing methods are also univariate. Examples are curve fitting and multicomponent analysis. If the signals of the compounds in the sample are overlapping or interfering, a fairly large amount of preinformation is required (Fig. 14) in order to find the solution. When such interferences are also occurring together with matrix effects, one has to calibrate with a Generalized Standard Addition Method (GSAM), developed by Kowalski [83]. GSAM is a combination of the standard addition method to correct for matrix effects, and a multicomponent analysis method to correct for interferences.

In some cases, many different spectra (or chromatograms) of the same object are available. For inhomogenous objects, for example, several samples of different constitution can be taken. This allows to apply multivariate data processing techniques. When the signals of the compounds in the sample are specific and linearly additive, the number of compounds which contribute to the signal, can be determined by a Principal Components Analysis (PCA) [84] (see Sect. 3.2.1). Without knowing the identity of *all* compounds, which are present and without knowing their spectra, a calibration by partical least squares (PLC) [85] allows to quantify the compounds of interest.

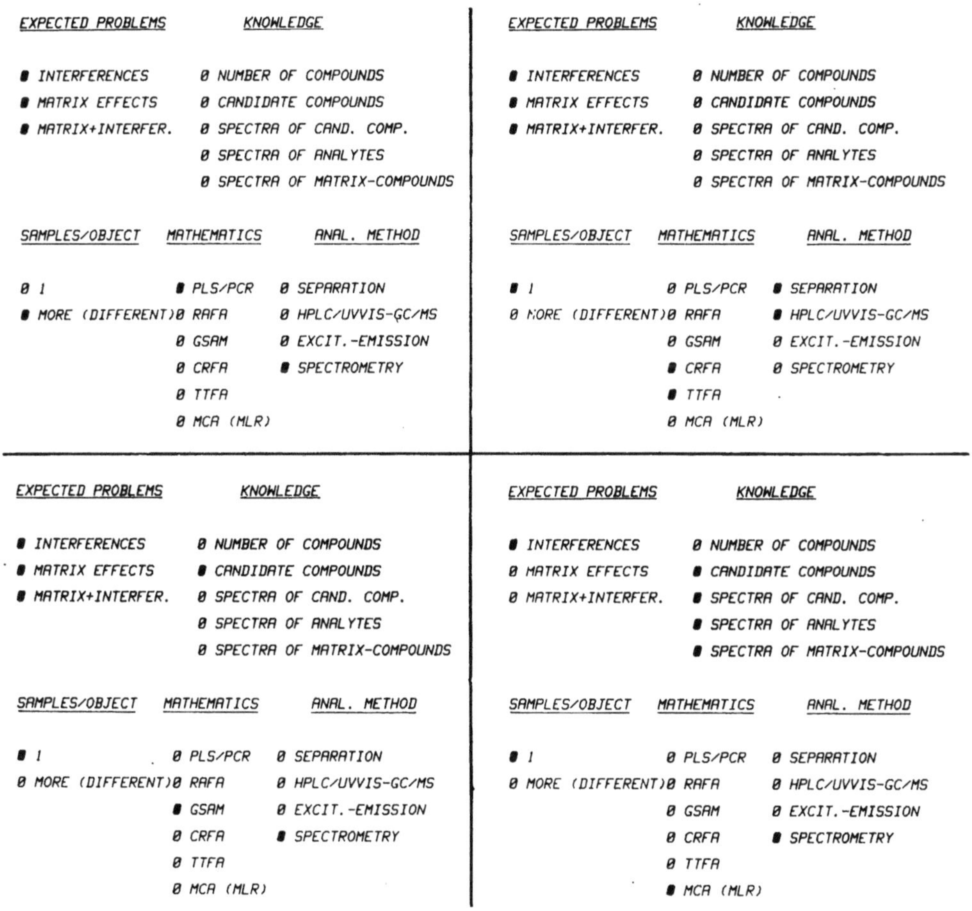

Fig. 14. Type of analytical method and Chemometrics required to analyse mixtures

The only condition for the application of PLS is that several samples are available with known amounts of the compounds of interest, for calibration. Interferences and matrix effects of the unknown compounds have not to be taken into account and do not effect the accurary of the analytical result (see further Section 3.3). The presence of candidate compounds, can be confirmed one at a time by a Target Transformation Factor analysis (TTFA) [86].

Above mentioned examples clearly show that if multivariate data processing methods are applicable, analytical information can be derived with a minimal amount of pre-information and a foreseeing of a maximum of problems. When the sampled object is homogenous, multivariate methods are only applicable when the analytical method itself produces multivariate signals. This is the case when several signals (e.g. spectra) are obtained for the sample as a function of another variable (e.g. time, excitation wavelength). For example in GC-MS, a mass spectrum is measured of the eluents every .1 à 1 second. In excitation-emission spectroscopy, spectra are measured at several excitation-wavelengths. The potentials of the application of multivariate statistics to such cases are very well demonstrated by the results obtainable by curve resolution factor analysis in HPLC with a UV-VIS photodiode array detector. Also when no pre-knowledge is available on the number and the identity of the compounds, pure spectra and elution profiles can be calculated for partially separated mixtures [87, 88].

In the paragraphs below, some of the above mentioned multivariate methods will be discussed in somewhat more detail, with respect to the data processing of signals obtained for hyphenated methods of the type: chromatography-spectrometry and spectrometry-spectrometry.

3.2.1 Chromatography-Spectrometry

Many applications of GC-MS and HPLC-UV-VIS prove the powerful capabilities of these methods to analyze complex mixtures. The principal limiting factor is the obtainable separation, which can be optimized as described in section 3.1. When the physico-chemical separation is incomplete, a mathematical improvement of the resolution can be considered by the application of multivariate statistics.

In chromatography-spectrometry the measured data are arranged in a data matrix: **D**. The rows of **D** represent spectra and the columns are chromatograms. When **S** is the datamatrix of the pure spectra of the compounds, then **D** can be decomposed into: $\mathbf{D} = \mathbf{C} \cdot \mathbf{S}$, where **C** is the matrix of the concentrations of all compounds in the various spectra.

Thus: **D** is a matrix of NS spectra by NW wavelengths
C is a matrix of NS times by NC concentrations
S is a matrix of NC pure spectra by NW wavelengths

In the worst case, **C**, **S** and the number of compounds are not known. Malinowski [84, 89] has extensively studied the applicability of principal components analysis (PCA) to determine the number of components in a data matrix. A basic explanation of PCA can be found in [34] and is briefly repeated below:

Any datamatrix **D**, can be decomposed in two other matrices, **A** and $\mathbf{V}_q$, where $\mathbf{D} = \mathbf{A} \cdot \mathbf{V}_q + \mathbf{E}$, $\mathbf{V}_q$ is a NC × NW matrix, and **A** is a NS × NC matrix with $\mathbf{V}_q$ having the following properties: the internal product of any pair of rows of $\mathbf{V}_q$ is zero (orthogonality) and NC is the smallest number of rows in $\mathbf{V}_q$, necessary to represent **D** within the noise (**E**). The rows of $\mathbf{V}_q$ are the first eigenvectors of the variance-covariance matrix of **D**. Because the dimensions of matrices **S** and $\mathbf{V}_q$ are equal, one can consider $\mathbf{V}_q$ as being a set of "abstract" spectra.

An alternative decomposition of the data matrix **D** is: $\mathbf{D}^T = \mathbf{B} \cdot \mathbf{V}_r + \mathbf{E}$, in which $\mathbf{V}_r$ has the same properties as $\mathbf{V}_q$, but represents by now "abstract" elution profiles. The rows of $\mathbf{V}_r$ are the first NC eigenvectors of the variance covariance matrix of $\mathbf{D}^T$.

Several methods have been proposed to determine the minimal number of rows, NC, in $\mathbf{V}_q$ or $\mathbf{V}_r$. When **D** is a data matrix of linearly additive signals, then NC represents the number of chemical species that contribute to the total signal. An example of the decomposition of a HPLC-UV-VIS data matrix in "abstract" elution profiles is shown in Fig. 15. These "abstract" elution profiles and "abstract" spectra can not directly be interpreted into qualitative and quantitative information. For that purpose it is necessary to transform $\mathbf{A} \cdot \mathbf{V}_q$ into $\mathbf{C} \cdot \mathbf{S}$, or to transform $\mathbf{B} \cdot \mathbf{V}_r$ into $\mathbf{S}^T \cdot \mathbf{C}^T$. In

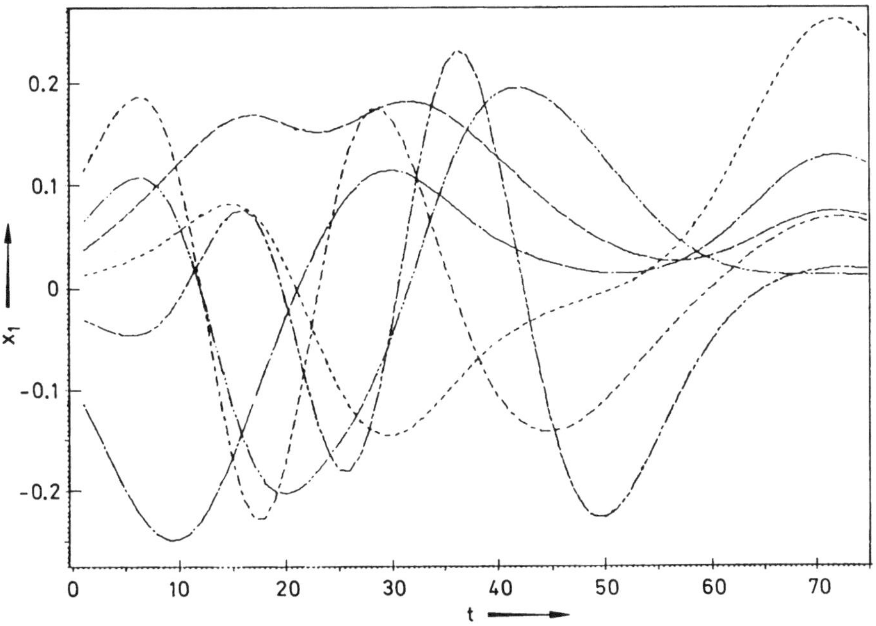

Fig. 15. "Abstract" elution profiles, represented by the eigenvectors of the variance-covariance metrix of a HPLC-UV-VIS data matrix of spectra. From B. G. M. Vandeginste, W. Derks, G. Kateman, Anal. Chim. Acta *173*, 259 (1985). Reproduced by permission of Elsevier Science Publishers, Amsterdam

other words the NC rows in $\mathbf{V}_q$ have to be transformed into NC pure spectra or the NC rows in $\mathbf{V}_r$ have to be transformed into NC elution profiles. This operation is Factor analysis. In principle there are an infinite number of transformations applicable on $\mathbf{V}_q$ and $\mathbf{V}_r$. Such a transform can be represented as a rotation matrix, which rotates $\mathbf{V}_q$ into $\mathbf{S}$ or $\mathbf{V}_r$ into $\mathbf{C}^T$, namely:

the pure spectra are $\mathbf{S} = \mathbf{R}_q \cdot \mathbf{V}_q$

the pure elution profiles are: $\mathbf{C}^T = \mathbf{R}_r \cdot \mathbf{V}_r$.

The key problem in factor analysis is to find a good transformation matrix $\mathbf{R}_q$ or

$\mathbf{R}_r$. Two approaches have been successfully applied: $\mathbf{R}_q$ and $\mathbf{R}_r$ are calculated by imposing constraints on the solutions $\mathbf{S}$ and $\mathbf{C}^T$. This method is called curve resolution factor analysis [88]. The second approach is based on the fact that candidate pure spectra or pure elution profiles are available. Such a candidate is called a Target. If a target (T) is indeed one of the pure spectra or one of the elution profiles, it should be one of the rows of the matrix $\mathbf{S}$ or $\mathbf{C}^T$. Knowing $\mathbf{V}_q$ respectively $\mathbf{V}_r$, one can calculate the corresponding vector, which transforms the abstract spectra or abstract elution profiles into the Target, which is being tested. A check whether the tested Target was indeed one of the rows in $\mathbf{S}$ or $\mathbf{C}^T$ is obtained by testing whether the calculated vector $\mathbf{R}_T$, transforms $\mathbf{V}_q$ or $\mathbf{V}_r$ back into the Target, $\mathbf{T}$ or by testing $\mathbf{R}_T.\ \mathbf{V} = \hat{\mathbf{T}} \overset{?}{=} \mathrm{T}$. If not, the target is marked invalid. Otherwise, the target was one of the true factors. This

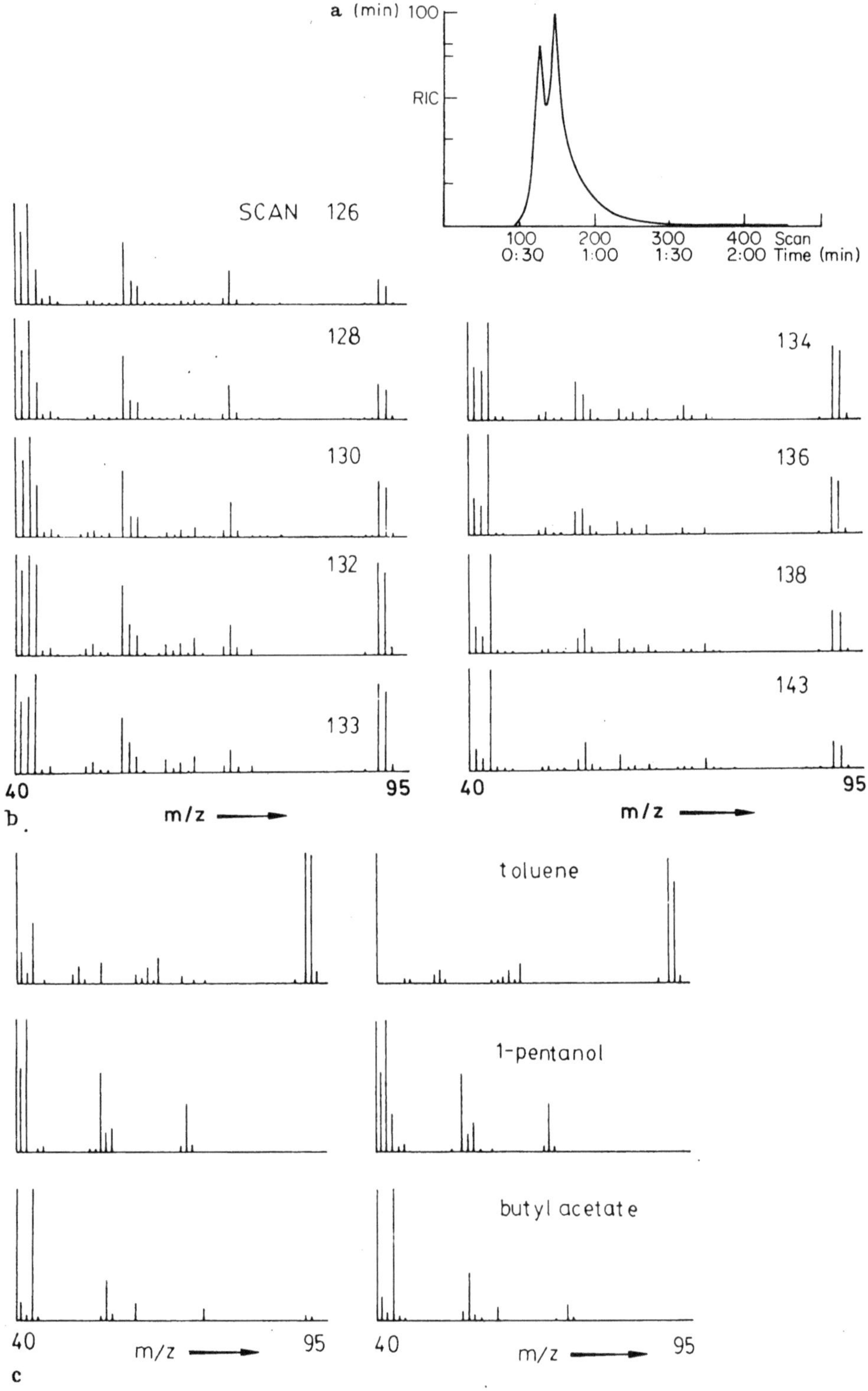
a (min) 100
RIC
100 0:30
200 1:00
300 1:30
400 2:00
Scan
Time (min)
SCAN 126
128
130
132
133
134
136
138
143
40
95
b
m/z
toluene
1-pentanol
butyl acetate
m/z
c

method is called Target Transformation Factor Analysis [86]. Lawton and Sylvestre [90] suggested in 1971 that spectra could be very well estimated by applying a minimum of simplifying assumptions; namely, all spectra and elution profiles should be non-negative. Unfortunately, only solutions could be obtained for systems with 2 components. Kowalski picked up the method and demonstrated its usefulness in LC-UV [91] and GC-MS [87,92]. Chen [93] and Vandeginste [88] extended the method to the three component case by adding more specific chemical knowledge. Chen [93] assumed that in GC-MS the pure spectra should be as simple as possible (smallest number of peaks). Vandeginste [88] added specific chromatographic knowledge by selecting the elution profiles with the smallest possible band width. Figure 16 shows the results obtained by Chen [93] for the GC-MS analysis of a mixture of 1-pentanol, toluene and butyl-acetate. The compounds are only partially resolved (Fig. 16a). It is seen that the mass spectra for the scan numbers 126 to 143 are mixture spectra (Fig. 16b). By the application of curve resolution factor analysis, three mass spectra could be calculated, which resemble very well the true pure mass spectra of the three compounds (Fig. 16c). Because no mathematical model is assumed for the elution profiles, these above mentioned methods are called self-modeling. The assumptions formulated by Chen [93] and Vandeginste [88], however, are insufficient to solve systems with more than three components, by self-modeling curve resolution factor analysis. More specific knowledge of the system has to be included in the algorithm with the danger of becoming less universally valid.

Knorr [94] and Frans [95] proposed to include a mathematical model for the shape of the elution profiles and could resolve systems with 6 and more overlapping peaks. They also demonstrated that the method is fairly robust for the inaccuracy of the models, describing the shape of the elution profiles. Independently from each other, Gemperline [96] and Vandeginste [97] argued that the attractive feature of self-modeling can be kept for the resolution of systems with more than three components by the application of an iterative version of Target Transformation Factor Analysis (ITTFA). ITTFA was first developed by Roscoe [98] and Hopke [99] for the factor analysis of the element concentrations in dust, and was successfully applied to the factor analysis of the element concentrations found at the intersection of two crossing lavabeds [99]. In ITTFA, an invalid Target is modified to a possibly better Target, which is thereafter resubmitted to the Target transformation factor analysis. In the particular application of LC-UV, the Target is a guess of one of the elution profiles. When the test $\hat{\mathbf{T}} = \mathbf{T}$ fails, the calculated target $\hat{\mathbf{T}}$ is modified: e.g. all negative values are set zero and all shoulders are removed. Gemperline [96] and Vandeginste [97] reported that the iterative process indeed converges to a stable and valid solution. Good results were obtained for 4, 6 and 8 component systems without assuming any peak shape. Figure 17 shows the resolution of a 6 component system and the convergence. By now LC-UV

◀ **Fig. 16a–c.** Resolution of GC-MS spectra obtained for mixtures of 1-pentanol, toluene and butyl-acetate. **a**. the elution profile. **b**. the mass spectra measured for scans 126 to 143. **c**. the calculated pure mass spectra from scans 126 to 143 compared to the pure mass spectra. From J. H. Chen and L. P. Hwang, Anal. Chim. Acta *133*, 277 (1981). Reproduced by permission of Elsevier Science Publishers, Amsterdam

instruments are becoming available in which capabilities for curve resolution factor analysis are present.

Similar results have been reported for LC-IR systems [100,101].

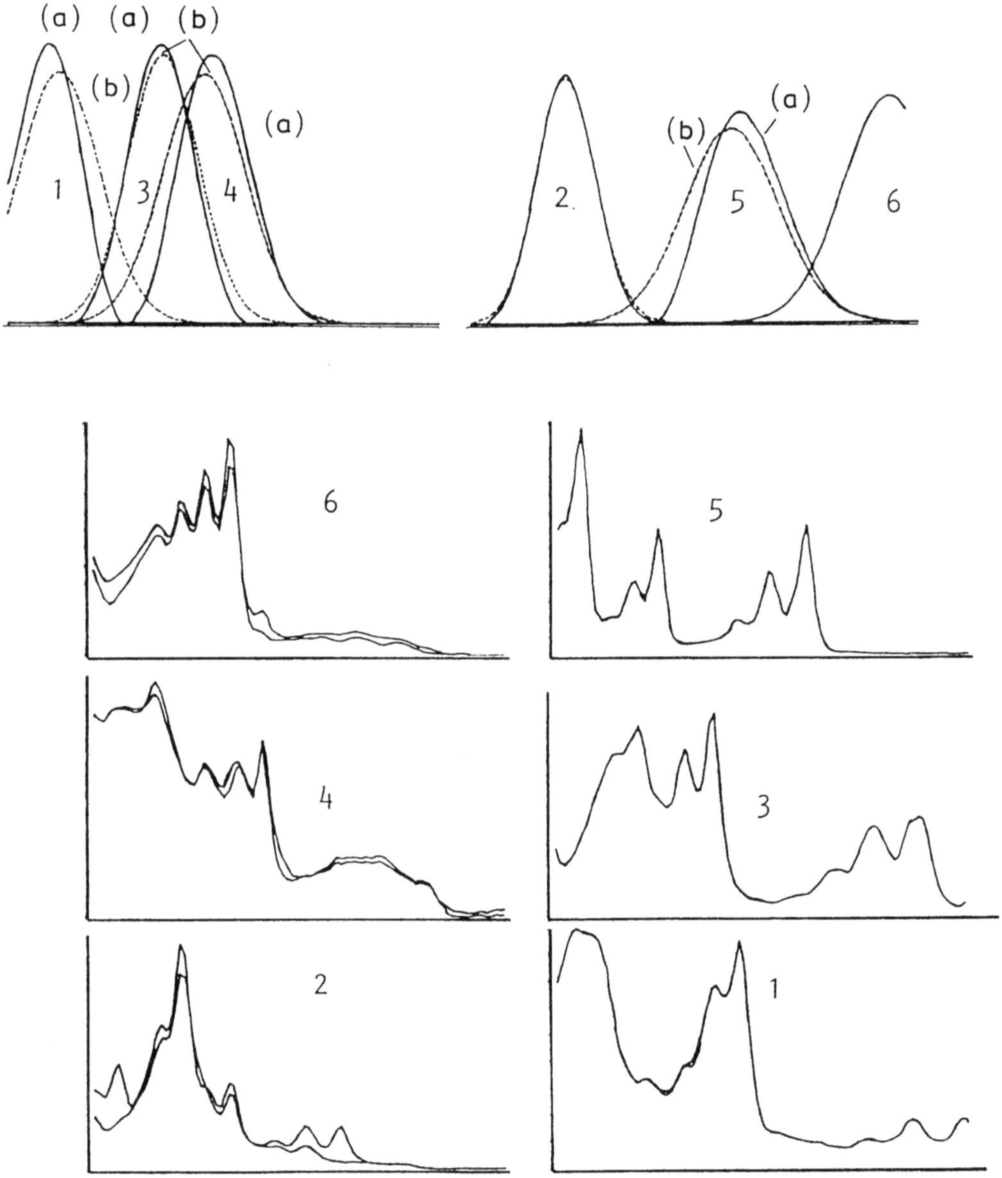

Fig. 17. The resolution of a HPLC-UV-VIS datamatrix obtained for a 6-component system by Iterative Target Transformation Analysis (ITTFA). [a] calculated elution profile, [b] true elution profile. From B. G. M. Vandeginste, W. Derks and G. Kateman, Anal. Chim. Acta *173*, 261 (1985). Reproduced by permission of Elsevier Science Publishers, Amsterdam

3.2.2 Spectrometry-Spectrometry

An important hyphenated method, which combines two spectrometric measuring principles is excitation-emission spectrometry (EES), the application of EES has been very much simplified by the introduction of the video-fluorimeter [102], which is capable of recording fluorescence spectra in a short time (0.01–0.001 sec) at a large number of excitation wavelengths. The measured datapoints are arranged in a so-called excitation-emission matrix (EEM).

The processing of EEM-data, obtained for mixtures, consists of the determination of the number of components, their concentrations and pure spectra. A very powerful method in that respect is rank annihilation factor analysis (RAFA). Based on RAFA, Ho et al. [103,104], derived a quantitative method for the determination of one or more compounds in a mixture, of which not all components are known. The rank of an EEM matrix theoretically is equal to the number of components in the mixture. Thus the rank of an EEM-matrix of a pure standard (or compound) should be equal to one.

Under ideal noise-free conditions one can diminish the rank (n) of a measured EEM-datamatrix of n-compounds, $\mathbf{M}$, by one, by subtracting the exact amount of EEM, $\mathbf{M}_K$, measured for the standard, K.
Thus Thus

$$\mathbf{N} = \mathbf{M} - a_k \cdot \mathbf{M}_k$$

where $\mathbf{N}$ is the matrix from which compound, k, has been removed, and a_k is the ratio between the concentration of compound k in the sample (c_k), and the standard (C_k^0): $a_k = C_k/C_k^0$.
The rank of a matrix is the number of significant eigenvectors, necessary to represent the datamatrix (see 3.2.1.). The value a_k is calculated by evaluating the eigenvalue of the least significant eigenvector as a function of a_k. For an increasing value of a_k, starting from zero, the least significant eigenvalue diminishes first because the component is being removed. After having passed through a minimum, this eigenvalue increases, because the compound has been overcompensated and introduces a negative concentration. The exact concentration a_K is found at the minimum. This is shown in Fig. 18, in which the eigenvalue of the least significant eigenvector is given as a function of a_K, by the removal of the spectrum of ethylbenzene [105] from the EEM spectrum of a mixture of ethylbenzene, o-xylene and p-xylene.

The method was applied on a 6-component mixture of polycyclic aromates. Each compound was quantified, independently from the presence of other compounds, and within the error of preparing standards [105]. It means that the method is fully capable to handle unknown and uncorrected interferences.

Because the structure of EEM is very similar to a datamatrix, measured by LC-UV, one could presume that the curve resolution factor analysis, described in 3.2.1. could also be useful. There is, however, one important difference: the only applicable constraint here is that the excitation-emission spectra are non-negative. Other constraints, which are valid in HPLC, e.g. that the elution profile with the smallest bandwidth should be selected, are missing here. For this reason curve resolution factor analysis, applied on EEM to find the pure excitation and emission spectra, can only

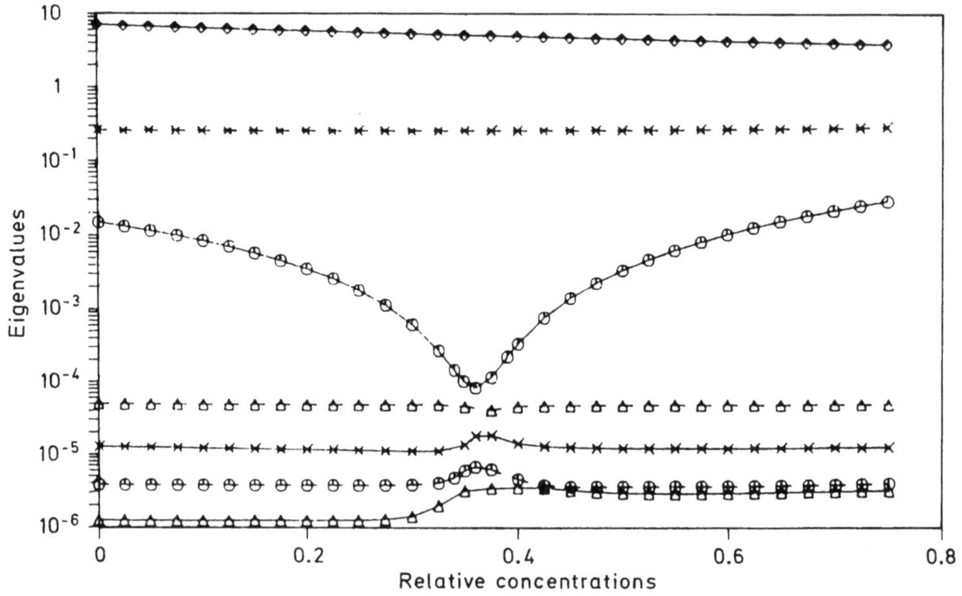

Fig. 18. The eigenvalues of an excitation-emission datamatrix (EEM) of a 3-component mixture of polycyclic aromates, as a function of the factor a_K, with which the EEM of ethyl benzene is removed. From M. McCue and E. R. Malinowski, J. Chromatogr. Sci. *21*, 233 (1983). Reproduced by permission of Preston Publications Inc., Niles, Illinois

solve binary mixtures [106]. A scheme which gives the accuracy of the estimated spectra, given as a function of spectral overlap can be found in [106].

Because emission spectra are strongly correlated with the excitation spectra the EEM datamatrix does not contain much qualitative information in addition to the regular absorbance spectrum. If, however, a videofluorimeter (VF) is coupled to a HPLC, a powerful device is obtained, which combines three measuring principles: separation-excitation-emission. Such a type of instrument, described by Hershberger [107], produces a three-dimensional datamatrix. The data analysis of an unknown mixture consists then of the following points: the determination of the number of compounds, the excitation-emission spectra of the compounds and the elution profiles. A three-dimensional data treatment has been developed by Appelloff and Davidson [108, 109]. They found that by the application of abstract factor analysis, an exact solution for above mentioned problems can be obtained for binary, ternary and quarternary mixtures. As a next step, after having calculated the pure spectra, they applied Rank annihilation factor analysis (RAFA) to quantify the amounts of the compounds in the mixture [109]. Conclusion is that HPLC-VF is capable to provide all quantitative and qualitative information without having any knowledge on the number and identity of the compounds present and with a minimum of chromatographic separation.

3.3 Multivariate Calibration

Because so many factors determine the response obtained for a chemical substance in a sample, it is usually not possible to derive directly the concentration from the measured response. The relationship between signal, or response and concentration has to be determined experimentally, a step which is called calibration. The complexity of the calibration depends upon the type of expected problems. These are roughly divided into three categories: interferences, matrix effects or interactions and a combination of both, a so-called interacting interference.

An interferent is a compound, other than the analyte, which contributes to the measured signal. When a compound other than the analyte does not contribute to signal, but instead influences the relationship between the analyte concentration and signal, there is a matrix interaction. An interacting interferent does both: it contributes to the signal and changes the sensitivity for the analyte.

Since a long time analytical chemists correct for interferences by a multicomponent analysis (MCA) and correct for interactions by a standard addition (SA) method. These methods are discussed in every elementary textbook on analytical chemistry. In MCA, the system is calibrated with pure standards. A problem encountered here is that all interferents have to be known, to be determined and to be corrected for. In SA the calibration is carried out in the sample and is repeated for each sample. A condition for a proper calibration by SA is that interferents should be absolutely absent. In practice, however, conditions for the application of MCA or SA are frequently not fulfilled, because of the presence of interacting interferents. The solution to that problem is usually sought along chemical paths: extraction, separation, masking agents. As a consequence analysis times tend to become too long and to require much manipulations to allow an easy automation. In addition precision is sometimes lower. A mathematical solution is, therefore, more desirable. From the fact that one should calibrate at a number of sensors (e.g. wavelengths) which is at least equal to the number of analytes plus interferents, it is obvious that the mathematical solution will be a multivariate one. The kind of mathematics involved depends upon the type of available standards to establish the relationship between concentration and response. One can directly observe that relationship by using pure standard solutions, which is called direct calibration, or direct multicomponent analysis [110,111]. When that relationship is implicitely observed on mixtures of known constitution, the calibration is indirect. For both types of calibration, chemometricians have developed and studied algorithms with which the chemical separation of interferents can be avoided. The two most important are the Generalized Standard Addition Method (GSAM) for direct calibration, developed by Saxberg and Kowalski [112], and Partial Least Squares (PLS), for indirect calibration developed by Wold and Martens [113,114]. Because these chemometric tools are relatively new, they are discussed in somewhat more detail, below.

GSAM is a combination of the standard addition method and multicomponent analysis. The general calibration model in GSAM is $\mathbf{R}_0 = \mathbf{K} \cdot \mathbf{C}_0$, where $\mathbf{R}_0$ is a vector of responses, obtained for the sample at NW sensors, $\mathbf{K}$ is the matrix of NW calibration factors of NA analytes, $\mathbf{C}_0$ is the vector of the concentrations of NA analytes, in the sample. The calibration of the system and the determination of the concentrations of the analytes are carried out in two steps.

First the responses $\mathbf{R}_0$ are measured for the sample. Thereafter $\mathbf{K}$ is determined by fitting the changes in the concentrations of the analytes in the sample, brought about by the standard additions, to the changes in the responses. Once all elements in the calibration matrix, $\mathbf{K}$, have been determined, the concentration vector $\mathbf{C}_0$ of the analytes in the sample is calculated. The method has been successfully applied to absorption spectrophotometry [83], anodic stripping voltametry [115] and ICP-atomic emission spectrophotometry [116]. Attractive features of the method are that automation is very easy [117] and automatic drift compensation is possible [118]. A drawback is that all interferents should be known and be corrected for.

For the application of indirect calibration methods, one should have several samples available with known constitution. These samples should be of exactly the same type as the unknown sample. For example for the determination of the fat, water and protein content in wheat by near infrared reflectance spectrometry, one should have available a number of wheat flour samples of which the amount of fat, water and protein is known, or determined by a conventional method.

The simplest form of multivariate calibration is multiple linear regression (MLR), which is well suitable in X-ray analysis [119]. The concentration of a given element is a linear combination of the intensities found at certain wavelengths in X-ray analysis.

A general model that includes most types of interactions can be written as [119]

$$\text{conc} = \sum_{j=1}^{n} b_j I_j + \sum_{j=1}^{n} \sum_{l=1}^{n} b_{jl} I_j I_l + b_0$$

where conc denotes the concentration of the component under study, I is the measured fluorescence intensity at the lines of the various compounds and b are the calibration constants.

In matrix notation:

$$\mathbf{C} = \mathbf{I} \cdot \mathbf{B}\,.$$

The calibration consists of determining all elements of the **B**-matrix, which requires as much samples as there are terms in the general model e.g. for a 3-component system Fe, Ni, Cr one can stipulate the following model for each of the elements

$$\begin{aligned}\text{conc} = {} & b_1 I_{Fe} + b_2 I_{Ni} + b_3 I_{Cr} + b_4 I_{Fe}^2 + b_5 I_{Fe} I_{Ni} + b_6 I_{Fe} IC_r \\ & + b_7 I_{Ni}^2 + b_8 I_{Ni} I_{Cr} + b_9 I_{Cr}^2 + b_{10}\end{aligned}$$

In this example, 30 calibration factors have to be quantified by the calibration procedure, which requires 30 independent equations and, therefore, 10 calibration samples (gives 3 equations per sample). Martens [120] warned against pitfalls by using MLR. There is, for example, no check whether the general model is adequate to describe the unknown samples. The calculated concentrations may be in error because the sample contains other impurities than present in the calibration samples, or because some external factors, such as the temperature, have disturbed the signal. Another problem is that it may be difficult to calculate the generalized inverse of matrix, **I**. By the application of principal components regression (PCR) [121] or partial least squares (PLS) instead of MCR, the pitfalls mentioned before can be avoided. The distinction between both methods is explained below.

Let us suppose that the concentrations of NA analytes are known in NS calibration standards. For each of the standards a spectrum is measured at NW wavelenghts. These spectra can be arranged in a datamatrix $\mathbf{D} = \mathbf{C} \cdot \mathbf{S}$, where **C** and **S** have the same meaning and dimensions as given in section 3.2.1. During the calibration step **D** is measured, and C is known. **S** can be determined by multiple linear regression. This requires to measure at least NA standards and to calculate the generalized inverse of the concentration matrix C, which in many instances is singular.

An alternative approach consists of first decomposing the measured datamatrix, **D**, into its principal components, $\mathbf{D} = \mathbf{A} \cdot \mathbf{V}_q + \mathbf{E}$ by applying the principles explained in section 3.2.1, $\mathbf{V}_q$ is the eigenvector matrix of **D**. The factor score matrix **A** can be calculated from **D** and $\mathbf{V}_q$ [88]. Thereafter the factor score matrix is related to the concentration of the NA analytes in the NS calibration samples by multiple linear regression, which gives: $\mathbf{C} = \mathbf{A} \cdot \mathbf{B}$, from which **B** can be calculated (**A** and **C** are known). From the dimensions of matrix **A** and **B** which are respectively (NS × NA) and (NA × NA), it follows that indirect calibration by PCR requires that the number of calibration samples is at least equal to the number of analytes (NS ≧ NA). The number of wavelenghts should also at least be equal to the number of analytes, but choosing a higher number has no effect on the number of required standards, allowing to use all relevant spectral information.

Other strong advantages of PCR over other methods of calibration are that the spectra of the analytes have not to be known, the number of compounds contributing to the signal have not to be known on the beforehand, and the kind and concentration of the interferents should not be known. If interferents are present, e.g. NI, then the principal components analysis of the matrix, **D**, will reveal that there are NC = NA + NI significant eigenvectors. As a consequence the dimension of the factor score matrix **A** becomes (NS × NC). Although there are NC components present in the samples, one can suffice to relate the concentrations of the NA analytes to the factor score matrix by $\mathbf{C} = \mathbf{A} \cdot \mathbf{B}$ and therefore, it is not necessary to know the concentrations of the interferents.

The concentrations of the analytes in an unknown sample are calculated from the measured spectrum, $\mathbf{d}_u$, as follows: first the factor scores, $\mathbf{a}_u$, of the spectrum of the unknown are calculated in the eigenvector space, $\mathbf{V}_q$, of the calibration standards: $\mathbf{a}_u = \mathbf{d}_u \cdot \mathbf{V}^T$. The concentrations of the NA analytes are calculated from the relationship found between the factor score and concentration:

$$\mathbf{C}_u = \mathbf{a}_u \cdot \mathbf{B}\,.$$

Combined in one step:

$$\mathbf{C}_u = \mathrm{d}_u \cdot \mathbf{V}_q^T \cdot \mathbf{B}\,.$$

In Partial Least Squares (PLS), which was introduced by Wold [113] in 1975, the matrices **D** and **C** are decomfosed:

namely, the measurement matrix, **D**, representing the spectra of the calibration standards and the concentration matrix, **C**, representing the concentrations of the analytes in the calibration standards, are decomposed by an iterative procedure, in such a way that the matrices **A** and **B** are equal:

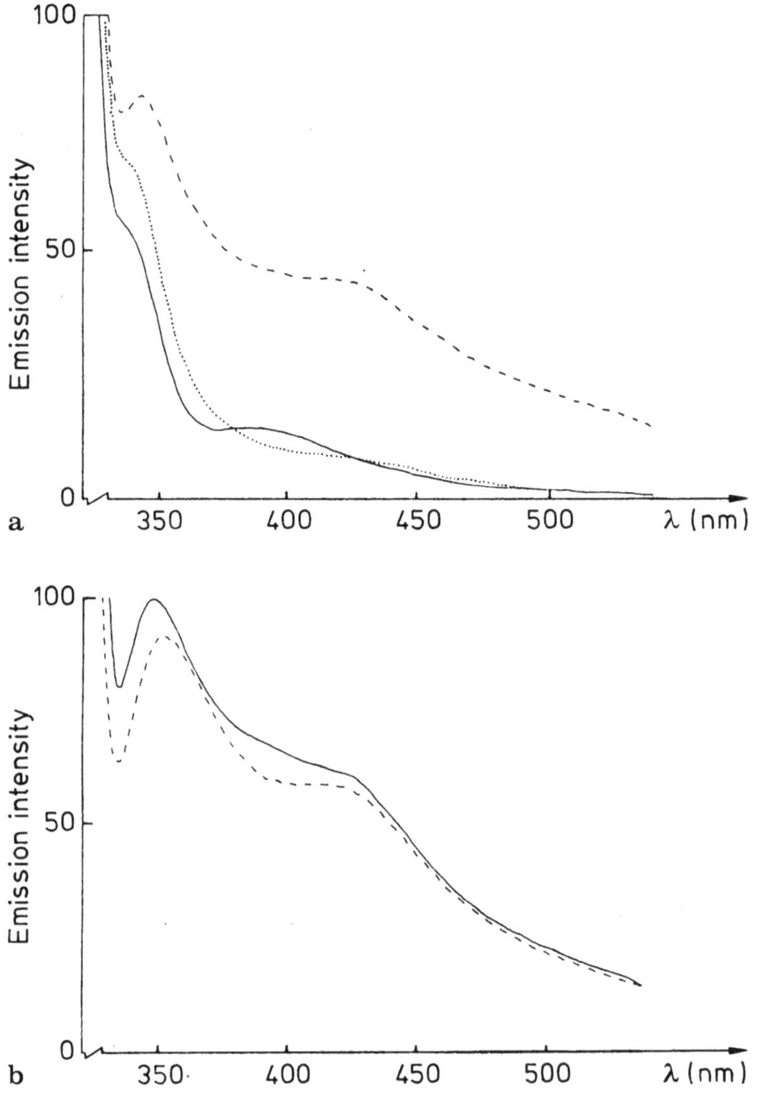

Fig. 19a and b. Spectra for the spectrofluorimetric analysis of a mixture of humic acid and ligninsulfonate, artificially contaminated with a whitenar. **a**. spectra of humic acid (------), lignin sulfonate (———) and a whitenar (·············); **b**. ——— expected spectrum for a mixture with absence of interactions ----------- measured spectrum. From W. Lindberg, J. A. Persson and S. Wold, Anal. Chem. *55*, 645 (1983). Reproduced by permission of the American Chemical Society, Washington DC

thus: $\mathbf{D} = \mathbf{A} \cdot \mathbf{V}_d + \mathbf{E}_d$ **D**: (NS × NW) matrix of the measurements
$\mathbf{C} = \mathbf{B} \cdot \mathbf{V}_C + \mathbf{E}_C$ **C**: (NS × NA) matrix of the concentrations
$\mathbf{V}_d$: (NC × NW) matrix of PLS loadings of the matrix **D**
$\mathbf{V}_C$: (NC × NA) matrix of PLS loadings of the matrix **C**
$\mathbf{A} \equiv \mathbf{B}$: (NS × NC) matrix of factor scores
NA: number of analytes
NC: number of compounds

A detailed description of the PLS matrix decomposition can be found in [85].

The concentrations of the analytes in an unknown sample are found from its spectrum d_u, by $\mathbf{C}_u = d_u \cdot (\mathbf{V}_d^T \cdot \mathbf{V}_C)$.

The potentials of the PLS algorithm are very well demonstrated on the spectrofluorimetric analysis of mixtures of humic acid and ligninsulfonate investigated by Lindberg et al. [85]. The problems associated with this analysis are the strong similarities between the spectra, without selective wavelengths and the interaction between the compounds, because of the difference between the emission spectra obtained of numerically added signals from pure substances and for a measured mixture of the same constitution (Fig. 19).

To demonstrate that PLS works in the case when the calibration is made with respect to a few "unknown" substances in a complex mixture, a calibration set was made with standards contaminated with an optical whitenar contained in a commercial detergent, having a strong emission spectrum in the measured 350–500 nm range. The concentration matrix (C) contains only the concentrations of ligninsulfonate and humic acid (table 10) in the calibration set. Thus the presence of the detergent was considered as

Table 10. Partial Least Squares (PLS) calibration and prediction of a complex mixture of humic acid and ligninsulfonate contaminated with an optical witenar. From W. Lindberg, J. A. Persson and S. Wold, Anal. Chem. *55*, 645 (1983). Reprinted by permission of the American Chemical Society, Washington DC

Compositions of the Calibration Solutions (μg mL^{-1})				
calibration set I		calibration set II		
humic acid	ligninsulfonate	humic acid	ligninsulfonate	detergent
1.24	0.111	3.01	0	0
4.12	0.099	0	0.401	0
1.52	0.492	0	0	90.6
3.05	0	1.48	0.158	40.0
0	0.390	1.12	0.410	30.4
2.25	0.148	3.40	0.303	50.8
2.33	0.219	2.43	0.298	70.6
1.34	0.134	4.02	0.115	89.4
0.931	0.263	2.28	0.504	81.8
3.68	0.312	0.959	0.145	101
1.40	0.415	3.19	0.253	120
2.94	0.305	4.13	0.569	118
3.50	0.341	2.16	0.436	27.6
3.02	0.671	3.09	0.247	61.7
		1.60	0.286	109
		3.16	0.701	60.0

an unknown disturbance. Table 11 shows that the concentrations are recovered for a broad concentration range of the "unknowns". One of the main advantages with the PLS model is that a criterion is obtained reflecting the similarity between samples and standards. In table 11 this is called a dissimilarity factor. A high dissimilarity factor means that other substances contaminate the sample, than present in the calibration set. Sample 18 in table 11 is such a case.

The PLS approach to multivariate linear regression modeling is relatively new and not yet fully investigated from a theoretical point of view. The results with calibrating complex samples in food analysis [122, 123] by near infrared reflectance spectroscopy, suggest that PLS could solve the general calibration problem in analytical chemistry.

Table 11. Results (μg mL^{-1}) for Mixtures with Three Substances Calibrated with Set II but Only Two Dependent Variables (Humic Acid and Ligninsulfonate)

sample no.	humic acid		ligninsulfonate		dissimilarity factor[b]
	actual concn	prediction error[a]	actual concn	prediction error[a]	
10	2.44	0.04	0.289	0.016	1.31
11	4.08	0.12	0.361	−0.100	3.51
12	1.06	0.04	0.234	0.044	0.75
13	3.32	0.07	0.123	−0.024	1.12
14	0.998	0.11	0.416	0.037	0.95
15	2.98	−0.10	0.403	0.010	1.55
16	5.13	−0.22	0.229	−0.109	2.20
17	5.06	−0	0	−0.017	1.80
18	0	−0.35	0.735	0.532	18.5
$\left(\frac{\Sigma(y_{true} - y_{calcd})^2}{n}\right)^{1/2}$		0.15		0.185	

4 New Directions in Chemometrics

During the last decade, analytical instrumentats became more fowerful but also more complex. A logical evolution from binary hyphenated methods to ternary (GC-FTIR-MS) and quarternary systems is obvious. It means that more advanced Chemometrics will be needed in the future to make sense of tomorrow's complex instrumentation (Kowalski [124]) and data. New technologies, such as robotics are being integrated into modern laboratory instrumentation. Also new software technologies, such as knowledge bases, empty shells for artificial intelligence, are becoming available. This could give the in pression that, Chemometrics is a trand "follower" by trying to adapt and exploit the advances in instrumentation and electronics. In my view Chemometrics should also play the role of an important trend setter. A result of Chemometric research e.g. is the development of algorithms which can handle "less prepared" or "dirty" samples (Martens [110, 111]). As a consequence the development of multisensory devices consisting of a large number of nonselective and nonspecific sensors becomes attractive because an adequate calibration is possible by now. Human interpretation of data structures measured with binary hyphenated methods is already quite difficult and could certainly refrain the development of ternary and quarternary systems, which would produce much non-consumable information. However, the fact that chemometric tools are available to derive all relevant qualitative and quantitative information from binary hyphenated methods, will certainly inspire instrument manufacturers to go ahead with these developments. A similar impact may be expected from the availability of sequential methods for optimization and calibration on the development of intelligent instruments. There are indications that methods from knowledge engineering could become of a growing importance to us, for method selection, method design, result interpretation, algorithm selection and laboratory management. Computers will be integrated in the analytical laboratory in the so-called computer integrated laboratory, consisting of a network of intelligent analytical workstations. An important aspect of an intelligent workstation is certainly the user-interface [125)], which may make computers easier to use for analytical chemists. The graphic-user interface is a comparatively recent development, which should enable analytical chemists to display analytical information and to interact with a computer in a user-friendly way. Display techniques based on image processing and pattern recognition will therefore, become a tool of an increasing importance. Pattern recognition, which is a very important topic of Chemometrics research, has been extensively studied in the past decades but remained difficult to be applied because of the large computer facilities required. A new trend is the release of microcomputer versions of packages for pattern recognition [126)] having an equal performance as the original mainframe version. The release of tested and refereed software by scientific publishers [127–133)] is a logical consequence of the demand for such software.

In qualitative analysis, structure elucidation by computer has been made possible. Three major approaches to chemical structure elucidation are: library searching, pattern recognition and artificial intelligence. At present the most pragmatically useful approach is based on library searching (LS) [134)]. LS features have been included in many mass and infrared spectrometers. A successful program for the interpretation of mass spectra, based on the methods from artificial intelligence, is DENDRAL [57)]. Future developments will certainly be directed towards an integrated approach of structure elucidation, in which pattern recognition, artificial intelligence and library search will be cooperatively applied to find "substructures" present in the compound using a combination of Mass, NMR and IR-spectral data. Furthermore, logics will become available which combines the substructures into a compound structure.

5 References

1. Massart, D. L.: Trends Anal. Chem. *1*, 348 (1982)
2. Christie, O.: lecture at the University of Birmingham, U.K., February 10, 1982
3. Wangersky, P. J.: Trends Anal. Chem. *1*, 150 (1982)
4. Leemans, F. A.: Anal. Chem. *43* (11), 36A (1971)
5. Currie, L. A., Filliben, J. J., DeVoe, J. R.: ibid. *44*, 497R (1972)
6. Schoenfeld, P. S., DeVoe, J. R.: ibid. *48*, 403R (1976)
7. Eckschlager, K., Stepanek, V.: ibid. *54*, 1115A (1982)
8. Müskens, P. J. W. M.: Dutch Ph.D. Thesis (1978)
9. Acland, J. D., Lipton, S.: J. Clin. Pathol. *24*, 369 (1971)

10. Frank, I. E., Kowalski, B. R.: Anal. Chem. *54*, 232R (1982)
11. Kowalski, B. R.: Chem. Ind. (London), *22*, 882 (1978)
12. Chemometric Newsletter, no. 12, July (1985)
13. Tuinstra, J., V. d. Vinne, J., Doornbos, D. A.: Trends Anal. Chem. *2* (10), V, (1983)
14. Betteridge, J.: lecture presented at the Symposium "The integrated approach to laboratory automation", RCS, Analytical Division, Dorset, October 1985
15. Cooley, J. W., Tukey, J. W.: Math. of Comput. *19*, 297 (1965)
16. Savitzky, A., Golay, M. J. E.: Anal. Chem. *36*, 1627 (1964)
17. Jones, R. N. et al.: Nat. Res. Counc. Can. Bull. *11* (1968)
18. Jones, R. N. et al.: ibid. *12* (1968)
19. Jones, R. N. et al.: ibid. *13* (1969)
20. Hirschfeld, T.: Anal. Chem. *52*, 297A (1980)
21. Jurs, P. C., Kowalski, B. R., Isenhour, T. L., Reilley, C. N.: ibid. *41*, 690 (1969)
22. Jurs, P. C., Kowalski, B. R., Isenhour, T. L., Reilley, C. N.: ibid. *41*, 1949 (1969)
23. Kowalski, B. R., Jurs, P. C., Isenhour, T. L., Reilley, C. N.: ibid. *41*, 695 (1969)
24. Spendley, W., Hext, G. R., Himsworth, F. R.: Technometrics *4*, 441 (1962)
25. Long, D. E.: Anal. Chim. Acta *46*, 193 (1969)
26. Deming, S. N., Parker, jr., L. R.: CRC Crit. Rev. Anal. Chem. *7*, 187 (1978)
27. Gottschalk, G.: Z. Anal. Chem. *258*, 8 (1972)
28. Malissa, H., Jellinek, G.: ibid. *289*, 1 (1978)
29. Dohmen, F. T. M., Thijssen, P. C.: Trends Anal. Chem. *4*, 167 (1985)
30. Eckschlager, K.: Z. Anal. Chem. *277*, 1 (1975)
31. Eskes, A., Dupuis, P. F., Dijkstra, A., De Clercq, H., Massart, D. L.: Anal. Chem. *47*, 2168 (1975)
32. Dupuis, F., Dijkstra, A.: ibid. *47*, 379 (1975)
33. Van Marlen, G., Dijkstra, A.: ibid. *48*, 595 (1976)
34. Massart, D. L., Dijkstra, A., Kaufman, L.: in "Evaluation and optimization of laboratory methods and analytical procedures", Elsevier, Amsterdam, 1980, Chapters 21 and 22
35. Vandeginste, B. G. M.: Anal. Acta *122*, 435 (1980)
36. Ackoff, R. L., Sasieni, M. W.: in "Fundamentals of operations research", Wiley, New York 1968
37. Massart, D. L., Janssens, C., Kaufman, L., Smits, R.: Anal. Chem. *44*, 2390 (1972)
38. De Clercq, H., Despontin, M., Kaufman, L., Massart, D. L.: J. Chromatogr. *122*, 535 (1976)
39. Massart, D. L., Janssens, C., Kaufman, L., Smits, R.: Z. Anal. Chem. *264*, 273 (1973)
40. Vandeginste, B. G. M.: Anal. Chim. Acta *112*, 253 (1979)
41. Kowalski, B. R.: Anal. Chem. *50*, 1309A (1978)
42. Kowalski, B. R., ed.: "Chemometrics: theory and application", ACS Symposium, Series 52, American Chemical Society, Washington D.C. 1977
43. Harper, A. M., Duewer, D. L., Kowalski, B. R., Fashing, J. L.: in Kowalski, B. R., ed. "Chemometrics: theory and application", ACS Symposium, Series 52, American Chemical Society, Washington D.C., p. 14, 1977
44. ARTHUR, Infometrix, Inc., P.O. Box 25808, Seattle, Wa 98125, U.S.A.
45. Vandeginste, B. G. M.: Anal. Chim. Acta *150*, 199 (1983)
46. Vandeginste, B. G. M.: in Kowalski, B. R., ed. 'Chemometrics: mathematics and statistics in chemistry', NATO ASI, Series C, vol. 138, p. 467, 1983
47. Dessy, R. E.: Anal. Chem. *54*, 1167A (1983)
48. Dessy, R. E.: ibid. *54*, 1295A (1983)
49. Betteridge, D., Sly, T. J., Wade, A. P., Tillman, J. E. W.: ibid. *55*, 292 (1983)
50. Van der Wiel, P., Van Dongen, L. G., Vandeginste, B. G. M., Kateman, G.: Lab Microcomputer *2* (1), 9 (1983)
51. SUMMIT Chromatographic System, Bruker Spectrospin, Coventry, U.K.
52. TAMED Chromatographic System, LDC Milton Roy, Stone, U.K.
53. Thijssen, P. C., Wolfrum, S. M., Smit, H. C., Kateman, G.: Anal. Chim. Acta *156*, 87 (1984)
54. Thijssen, P. C., Prop, L. T. M., Kateman, G., Smit, H. C.: ibid. *174*, 27 (1985)
55. Rutan, S. C., Brown, S. D.: ibid. *167*, 23 (1985)
56. Rutan, S. C., Brown, S. D.: Anal. Chem. *55*, 1707 (1983)
57. Lindsay, R. K., Buchanan, B. G., Feigenbaum, E. A., Lederberg, H.: "Applications of artificial intelligence for organic chemistry. The DENDRAL project", McGraw-Hill, New York, 1984

58. Wipke, W. T., Brown, H., Smith, G., Choplin, F., Seber, W.: "SECS-Simulation and evaluation of chemical synthesis: strategy and planning", ACS 61, 1977
59. Corey, E. J., Wipke, W. T., Cramer, R. D., Howe, W. J.: J. Am. Chem. Soc. *94*, 421 (1972)
60. Derde, M. P., Massart, D. L.: Chemisch Magazine *40*, 563 (1984)
61. Hagewijn, G., Massart, D. L.: J. of Pharmaceutical and Biomedical analysis *1* (3), 331 (1983)
62. Dessy, R. E., (ed.): Anal. Chem. *56*, 1312A (1984)
63. Wade, A. P.: Ph.D. Thesis, University of Wales, Swansea (1985)
64. Vandeginste, B. G. M.: Analusis *12*, 496 (1984)
65. Klaessens, J. W. A., Kateman, G., and Vandeginste, B. G. M.: Trends Anal. Chem. *4* (5), 114 (1985)
66. Meglen, R.: in Chemometrics Newsletter *12*, 4 (1985)
67. Davies, O. L. (ed): "The design and analysis of industrial experiments", Oliver and Boyd, Edinburgh 1971
68. Morgan, S. L., Deming, S. N.: Anal. Chem. *46*, 1170 (1974)
70. Deming, S. N., Morgan, S. L.: ibid. *45*, 278A (1973)
71. Box, G. E. P., Draper, N. R.: "Evolutionary operation. A method for increasing industrial productivity", J. Wiley, New York 1969
72. Berridge, J. C.: "Techniques for the automated optimization of HPLC separations", J. Wiley, New York 1985
73. Morgan, S. L., Deming, S. N.: J. Chromatogr. *112*, 267 (1975)
74. Berridge, J. C.: ibid. *244*, 1 (1982)
75. Berridge, J. C., Morrissey, E. G.: ibid. *316*, 69 (1984)
76. Glajch, J. L., Kikrkland, J. J., Squire, K. M., Minor, J. M.: ibid. *199*, 57 (1980)
77. Watson, M. W., Carr, P. W.: Anal. Chem. *51*, 1835 (1979)
78. Wegscheider, W., Lankmayr, E. P., Budna, K. W.: Chromatographia *15*, 498 (1982)
79. Nickel, J. H., Deming, S. N.: Liq. Chromatogr. Mag. *1*, 414 (1983)
80. Kester, A. S., Thompson, R. E.: J. Chromatogr. *310*, 372 (1984)
81. Dunn, D. L., Thompson, R. E.: ibid. *264*, 264 (1983)
82. Smits, R., Vanroelen, C., Massart, D. L.: Z. Anal. Chem. *273*, 1 (1975)
83. Jochum, C., Jochum, P., Kowalski, B. R.: Anal. Chem. *53*, 85 (1981)
84. Malinowski, E. R., Howery, D. G.: "Factor Analysis in Chemistry", Wiley, New York 1980
85. Lindberg, W., Persson, J. A., Wold, S.: Anal. Chem. *55*, 643 (1983)
86. Weiner, P. H., Malinowski, E. R., Levinstone, A.: J. Phys. Chem. *74*, 4537 (1970)
87. Sharaf, M. A., Kowalski, B. R.: Anal. Chem. *53*, 518 (1981)
88. Vandeginste, B. G. M., Essers, R., Bosman, T., Reijnen, J., Kateman, G.: ibid. *57*, 971 (1985)
89. Malinowski, E. R.: Anal. Chim. Acta *103*, 339 (1978)
90. Lawton, W. H., Sylvestre, E. A.: Technometr. *13*, 617 (1971)
91. Osten, D. W., Kowalski, B. R.: Anal. Chem. *56*, 991 (1984)
92. Sharaf, M. A., Kowalski, B. R.: ibid. *54*, 1291 (1982)
93. Chen, J. H., Hwang, L. P.: Anal. Chim. Acta *133*, 271 (1981)
94. Knorr, F. J., Torshim, H. R., Harris, J. M.: Anal. Chem. *53*, 821 (1981)
95. Frans, S. D., McConnell, M. L., Harris, J. M.: ibid. *57*, 1552 (1985)
96. Gemperline, P. J.: J. Chem. Inf. Comput. Sci. *24*, 206 (1984)
97. Vandeginste, B. G. M., Derks, W., Kateman, G.: Anal. Chim. Acta *173*, 253 (1985)
98. Roscoe, B. A., Hopke, P. K.: Comput. Chem. *5*, 1 (1981)
99. Hopke, P. K., Alpert, D. J., Roscoe, B. A.: ibid. *7*, 149 (1983)
100. Gilette, P. C., Lando, J. B., Koenig, J. L.: Appl. Spectrosc. *36*, 661 (1982)
101. Gilette, P. C., Lando, J. B., Koenig, J. L.: Anal. Chem. *55*, 630 (1983)
102. Warner, I. M., Callis, J. B., Davidson, E. R.: Anal. Lett. *8*, 665 (1975)
103. Ho, C. N., Christian, G. D., Davidson, E. R.: Anal. Chem. *50*, 1108 (1978)
104. Ho, C. N., Christian, G. D., Davidson, E. R.: ibid. *52*, 1071 (1980)
105. McCue, M., Malinowski, E. R.: J. Chromatogr. Sci. *21*, 229 (1983)
106. Warner, I. M., Christian, G. D., Davidson, E. R.: Anal. Chem. *43*, 564 (1977)
107. Hershberger, L. W., Callis, J. B., Christian, G. D.: ibid. *53*, 971 (1981)
108. Appellof, C. J. and Davidson, E. R.: ibid. *53*, 2053 (1981)
109. Appelof, C. J., Davidson, E. R.: Anal. Chim. Acta *146*, 9 (1983)
110. Martens, H., Naes, T.: Trends Anal. Chem. *3* (8), 204 (1984)

111. Naes, T., Martens, H.: ibid. *3* (10), 266 (1984)
112. Saxberg, B. E. H., Kowalski, B. R.: Anal. Chem. *51*, 1031 (1979)
113. Wold, H. in: "Perspectives in probability and statistics". Papers in honour of M. S. Bartlett (Gani, J. ed.), Academic Press, London 1975
114. Wold, S., Martens, H., Wold, H.: "The multivariate calibration problem in chemistry solved by the PLS method. Lecture notes in mathematics", Springer Verlag, Heidelberg, in press
115. Gerlach, R. W., Kowalski, B. R.: Anal. Chim. Acta *134*, 119 (1982)
116. Kalivas, J. H., Kowalski, B. R.: Anal. Chem. *53*, 2207 (1981)
117. Kalivas, J. H., Kowalski, B. R.: ibid. *55*, 532 (1983)
118. Kalivas, J. H., Kowalski, B. R.: ibid. *54*, 560 (1982)
119. Bos, M.: Anal. Chim. Acta *166*, 261 (1984)
120. Martens, H., Jensen, S. A.: "Partial least squares regression: A new two-stage NIR calibration method" in Proceedings of the 7-th World Cereal and Bread Congress, Prague, 1982, Elsevier, Amsterdam, in press
121. Mandel, J.: The American Statistician *36* (1), 15 (1982)
122. Sjostrom, M., Wold, S., Lindberg, W., Persson, J., Martens, H.: Anal. Chim. Acta *150*, 61 (1983)
123. Martens, H., Russwurm jr., H.: Food research and data analysis, Applied Science, Barking, Essex (U.K.) 1983
124. Kowalski, B. R., in Martens, H., Russwurm, jr. H.: Food research and data analysis, Applied Science, Barking, Essex (U.K.) 1983
125. O'Haver, T. C.: Trends Anal. Chem. *4* (8), 191 (1985)
126. In Sight: Infometrix, Inc., Seattle, Washington
127. Borman, S. A.: Anal. Chem. *57*, 983A (1985)
128. Kateman, G., Van der Wiel, P. F. A., Janse, T. A. H. M., Vandeginste, B. G. M.: "CLEOPATRA, Chemometrics library: an extendable set of programs as an aid in teaching, research and applications", Elsevier, Amsterdam 1985
129. Massart, D. L., Derde, M. P., Michotte, Y., Kaufman, L.: "BALANCE, a program to compare the means of two series of measurements", Elsevier, Amsterdam 1985
130. Deming, S. N., Morgan, S. L.: "INSTRUMENTUNE-UP: a program to assist the 'tune-up' of instruments', Elsevier, Amsterdam 1985
131. Baadenhuijsen, H., Arts, J., Somers, L., Smit, J. C.: "REFVALUE: a software package to calculate references intervals from total hospital patient laboratory data", Elsevier, Amsterdam 1985
132. Van der Wiel, P. F. A., Vandeginste, B. G. M., Kateman, G.: "COPS, Chemometrical optimization by Simplex", Elsevier, Amsterdam 1985
133. Thielemans, A., Derde, M. P., Rousseeuw, P., Kaufman, L., Massart, D. L.: "CLUE, a program for hierarchical divisive clustering", Elsevier, Amsterdam 1985
134. Hollowell, jr., J. R., Delaney, M. F.: Trends Anal. Chem. *4* (3), IV (1985)

Chemometrics — Sampling Strategies

Gerrit Kateman

Laboratory for Analytical Chemistry, Faculty of Sciences,
Catholic University of Nijmegen, Toernooiveld,
6525 ED Nijmegen, The Netherlands

Table of Contents

Sampling strategies depend on the type of the object to be sampled, the objective of the principal and the means that are available for analysis. Taking into account the properties of the object, optimal sampling schemes can be developed for particulate material and internally correlated objects (fluids, gases and soil). Depending on the objective these schemes are aimed at information optimization, cost minimization or optimal information for control. The analyzing laboratory has its influence on the information that can be obtained through its speed and capacity.

Most strategies can be quantified by chemometric techniques.

1 Introduction

Analytical chemistry is interested in information that can be obtained from material objects or systems. In more down to the earth terms this means that analytical chemists try to tell something new about objects, goods, bulk material or material systems. The way they obtain this information changes with the problem. In most cases they try to get the information from the qualitative or quantitative composition. A vast array of instruments and methods is to their disposition but most of them have one thing in common: their size is limited and the way they obtain information is destructive. That means that as a rule the analytical chemist cannot or will not use the whole object in his analysis machine, but that he uses only a small part of the object. In practice this fraction can be very small: the amount of material introduced in the analytical method rarely exceeds 1 g, but as a rule is not more than 0.01 to 0.1 g. This can be part of a shipload of ore, say 10^{11} g, or a river, transporting 10^{13}–10^{15} g water/day. In many instances this fraction is larger, but a fraction of the object to be analyzed of 10^{-4}–10^{-5} is common practice.

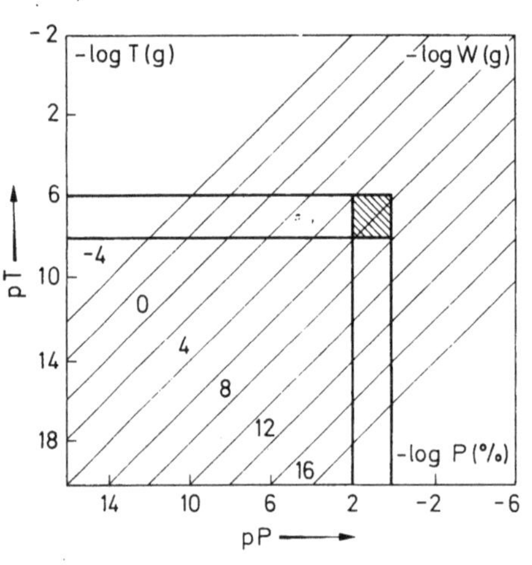

Fig. 1. Nomogram for the interdependence of sample size, composition, and total amount component [1]. Reprinted by permission of Springer Verlag from Arbeitskreis "Automation in der Analyse", Fres. Z. Anal. Chem. 261 (1972), p. 7

The implications of such a small fraction of the object to be investigated are enormous. The sample as we will call this fraction must fulfill a series of expectations before it may be called a sample and used as such. A sample must:

- represent the properties under investigation: composition, colour, crystal type, etc. faithfully;
- be of a size that can be handled by the sampler;
- be of a size than can be handled by the analyst, say from a 0.001–1.0 g;
- keep the properties the object had at the time of sampling, or change its properties in the same way as the object;
- be suitable to give the information the principal wants e.g. mean composition, composition as a function of time or place;
- keep its identity throughout the whole procedure of sampling, transport and analysis.

To satisfy these demands, the analyst can use the results of much theoretical and practical work from other disciplines.

Statistics and probability theory provided the analyst with the theoretical framework that predicts the uncertainties in estimating properties of populations when only a part of the population is available for investigation. Unfortunately this theory is not well suited for analytical sampling. Mathematical samples have no mass, do not segregate or detoriate, are cheap and are derived from populations with nicely modelled composition, e.g. a Gaussian distribution of independent items. In practice the analyst does not know the type of distribution of the composition, he has usually to do with correlations within the object and the sample or the number of samples must be small, as a sample or sampling is expensive.

Electronics uses mathematical sampling theories and extended it substantially. But electrons are light, cheap and abundant. Sampling rates can be very high in contrast to material sampling rates.

This isolated position of analytical sampling is probably the reason that sampling theory for analytical purposes developed late and slow. The art of constructing instruments for sampling and sample handling developed before theory, so many sampling instruments do not give good samples. Static statistical theories have been used for a long time to describe sampling as sampling from known, stationary, homogeneous populations. The abundance of papers describing these techniques is vast. In this review most of these papers are disregarded. Instead much emphasis has been laid on the attempts to describe theoretically the more complicated sampling problems, regarding more parameters than population parameters. For a more complete review of sampling, including sampling techniques, the reader should consult e.g. [2)].

2 Classification

To get an idea of the available knowledge about sampling, its theory and practice, a rough division will be made first of the different points of view of sampling.

As follows from the foregoing there can be distinguished three points of view:
— the principal and his objectives;
— the object and its properties;
— the analyst and his restrictions.

2.1 The Objective

The information that is wanted by the principal should comply with his objectives. Most information will be used for:
— describing the object globally or in detail;
— monitoring an object or system;
— controlling a system or process.

2.1.1 Description

Description of an object can be the determination of the gross composition for example, lots of a manufactured product, lots of raw material, single objects, the mean state of a process. Regarding sampling effort it is desirable to collect a sample that has the minimum size set by the condition of representativenes or demanded by handling.

Another goal can be the description of the object in detail, e.g. the composition of a metal part as a function of distance from the surface, or the composition of various particles in a mixed particulate product e.g. pigments. Here it is necessary to know the size and the number of samples, the distance between samples or the sampling frequency.

2.1.2 Monitoring

The next objective can be monitoring of an object or system as a function of time. Here it is often sufficient to know that a certain value of the property under investigation has been reached, or will be reached with a certain probability. If action is required it can be approximate.

Monitoring the effluent of a smoke stack, or the concentration of a drug in a patient requires that the sampling rate is as low as possible and that can be predicted with a known probability that between the sampling times no fatal concentration change will occur. If it is expected that the monitored value will exceed a preset value a simple action can prevent that: administer some drug, open or close a value, etc.

Another characteristic of threshold control is that in most cases only one level is monitored, the level pertaining to the high level threshold or to the lower level.

2.1.3 Control

The objective of control is quite different. The purpose of control is to keep a process property, e.g. the composition, as close to a preset value as is technically possible and economically desirable. The deviation from the set point is caused by intentional or random fluctuations of the process condition. In order to control the fluctuating process, samples must be taken with such frequency and analyzed with such reproducibility and speed that the process condition can be reconstructed. From this reconstruction predictions can be made for the near future and control action can be optimal. Another goal can be the detection of nonrandom deviations, like drift or cyclic variations. This also sets the conditions for sampling frequency and sample size.

2.2 The Object

In general only two types of objects can be distinguished: homogeneous and heterogeneous. It is clear that real homogeneous object do not pose much problems with regard to sampling. However, true homogeneous objects are rare and homogeneity may be assumed only after verification. Heterogeneous objects can be divided into two subsets, those with discrete changes of the properties and those with continuous changes. Examples of the first type are ore pellets, tablets, bulk blended fertilizer and coarse cristallized chemicals.

Examples of the other type are larger quantities of fluids and gases, including air, mixtures of reacting compounds and finely divided granular material.

Another description of heterogeneous objects can be: "Objects with random distribution of the property parameter". This type is rare and exists only when made intentionally. More common is the type that has a correlated distribution of random properties, e.g. the output of a chemical process or the composition of a flowing river. The behaviour of the properties can be understood by considering the process as a series of mixing tanks that connect input and output. The degree of correlation can be described by Fourier techniques or better by autocorrelation or semivariance.

A special type of heterogeneous objects exhibits cyclic changes of its properties. Frequencies of cyclic variations can be a sign of daily influences as temperature of the environment or shift-to-shift variations. Seasonal frequencies are common in environmental objects like air or surfacewater.

2.3 The Analyst

One important restriction set by the analyst and his instruments is already mentioned, the limited size of samples. Other restrictions influencing the act of sampling are the accuracy of the method of analysis, the limit of detection and the sensitivity. These analytical parameters set an upper limit on the information that can be obtained.

Another important analytical parameter is speed. The time between sampling and availability of the result affects the useable information for monitoring and control. The time required for analysis sets the time available for sampling. But there is another interrelation between sampler and analyst. The frequency of sampling causes a workload for the analyst and affects the analysis time by queueing of the samples on the laboratory bench, thus diminishing the information output. Obviously an optimal sampling frequency exists given the laboratory size and organization.

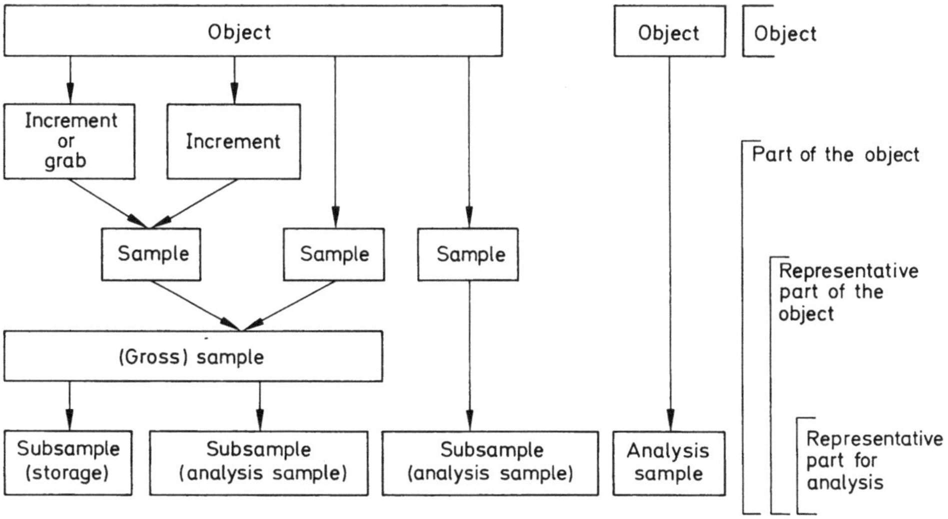

Fig. 2. Sample nomenclature [5]. Reprinted by permission of John Wiley & Sons, Inc. from G. Kateman and F. W. Pijpers, "Quality Control in Analytical Chemistry", p. 20, 1981. Copyright © 1981 John Wiley & Sons Inc

Table 1. Glossary of Terms Used in Sampling [2]

Bulksampling. Sampling of a material that does not consist of discrete, identifiable, constant units, but rather of arbitrary, irregular units.
Gross sample. (Also called bulk sample, lot sample.) One or more increments of material taken from a larger quantity (lot) of material for assay or record purposes.
Homogeneity. The degree to which a property or substance is randomly distributed throughout a material. Homogeneity depends on the size of the units under consideration. Thus a mixture of two minerals may be inhomogeneous at the molecular or atomic level but homogeneous at the particulate level.
Increment. An individual portion of material collected by a single operation of a sampling device, from parts of a lot separated in time or space. Increments may be either tested individually or combined (composited) and tested as a unit.
Individuals. Conceivable constituent parts of the population.
Laboratory sample. A sample, intended for testing or analysis, prepared from a gross sample or otherwise obtained. The laboratory sample must retain the composition of the gross sample. Often reduction in particle size is necessary in the course of reducing the quantity.
Lot. A quantity of bulk material of similar composition whose properties are under study.
Population. A generic term denoting any finite or infinite collection of individual things, objects, or events in the broadest concept; an aggregate determined by some property that distinguishes things that do and do not belong.
Reduction. The process of preparing one or more subsamples from a sample.
Sample. A portion of a population or lot. It may consist of an individual or groups of individuals.
Segment. A specifically demarked portion of a lot, either actual or hypothetical.
Strata. Segments of a lot that may vary with respect to the property under study.
Subsample. A portion taken from a sample. A laboratory sample may be a subsample of a gross sample; similarly, a test portion may be a subsample of a laboratory sample.
Test portion. (Also called specimen, test specimen, test unit, aliquot.) That quantity of material of proper size for measurement of the property of interest. Test portions may be taken from the gross sample directly, but often preliminary operations such as mixing or further reduction in particle size are necessary.

Ku [3] stated that a prerequisite to the development of an efficient analytical strategy is definition of the purpose for which the results are going to be used. This point has been laid down in the recommendation [4].

The foregoing introduction of sampling and influences on sampling has been treated in [5,6,7].

More general introductions in sampling and sampling statistics are given in [8,9,10,11].

Reviews on sampling are given in [12,13,2]. A review covering older references is e.g. [14].

Sampling error lecture demonstrations have been published by Bauer and Kratochvil [15,16]. A software program allowing simulation of many sampling situations and providing calculation algorithms for sampling schemes has been published [17].

Given the quite simple and clear model of sampling strategies and the economically very important impact of sampling there has been published comparatively little about sampling strategies. The emphasis has been more on analytical techniques. Detection limit, precision and capacity have been the main topics in analytical chemistry for more then 30 years. Chemometrics, providing means to extract more information from available data and tools to make experimental planning more efficient, added not much to sampling theory.

3 Sampling for Gross Description

3.1 Random Particulate Objects

The foundations for sampling theory, seen from the point of view of the object, have been laid by Benedetti-Pichler, Visman and Gy [18,19,20,21,22]. Their starting point was the object that consists of two or more particulate parts. The particulate composition is responsible for a discontinuous change in composition. A sample must be of sufficient size to represent the mean composition of the object. The cause of difference between an increment of insufficient size and the object is the inevitable statistical error. The composition of the sample, the collection of increments, is given by the mean composition of the object and the standard deviation of this mean. As the standard deviation depends on the number of particles, the size and the composition of these particles, an equation can be derived that gives the minimum number of particles.

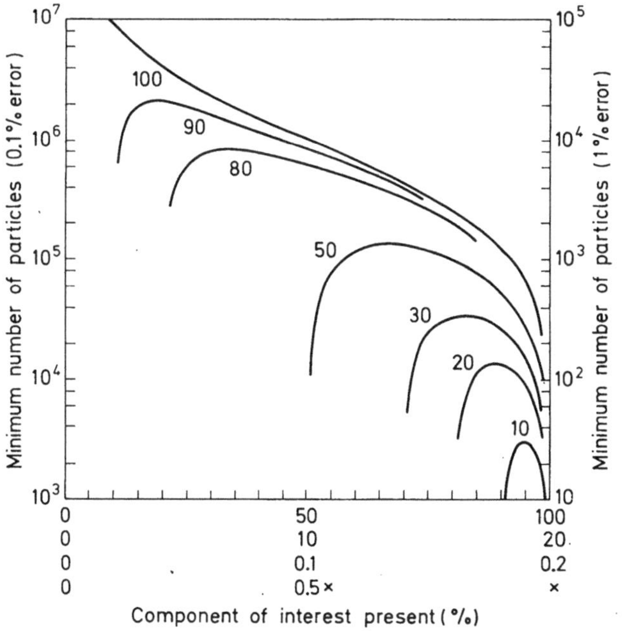

Fig. 3. Relation between the minimum number of units in a sample required for sampling errors (relative standard deviations in percentage) of 0.1 and 1% (y-axis) and the overall composition of a sample (x-axis), for mixtures having two types of particles with a relative difference in composition ranging from 100 to 10% [23]. Reprinted with permission from W. E. Harris, B. Kratochvil, Analytical Chemistry, *46* (1974), p. 314. Copyright 1974 American Chemical Society

Visman [19,20,21] described the sampling variance as the sum of random and segregation compounds according to $S^2 = (A/wn) + (B/n)$. Here A and B are constants determined by preliminary measurements on the system. The A term is the random component; its magnitude depends on the weight w and number n of sample increments collected as well as the size and variability of the composition of the units in the population. The B term is the segregation component; its magnitude depends only on the number of increments collected and on the heterogeneity of the population. Different ways of estimating the values of A and B are possible. Another approach to estimate the amount of samples that should be taken in a given increment so as not to exceed a predetermined level of sampling uncertainty is that through the use of Ingamells' sampling constant [24,25,26]. Based on the knowledge that the between-sample standard deviation decreases as the sample size is increased, Ingamells has shown that the relation $WR^2 = K_s$ is valid in many situations. W represents here the weight of the sample, R is the relative standard deviation (in percent) of the sample composition and K_s is the sampling constant, the weight of sample required to limit the sampling uncertainty to 1 % with 68 % confidence. The magnitude of K_s may be determined by estimating the standard deviation from a series of measurements of samples of weight W. An example of an Ingamells sampling constant diagram is shown in Fig. 4 [27].

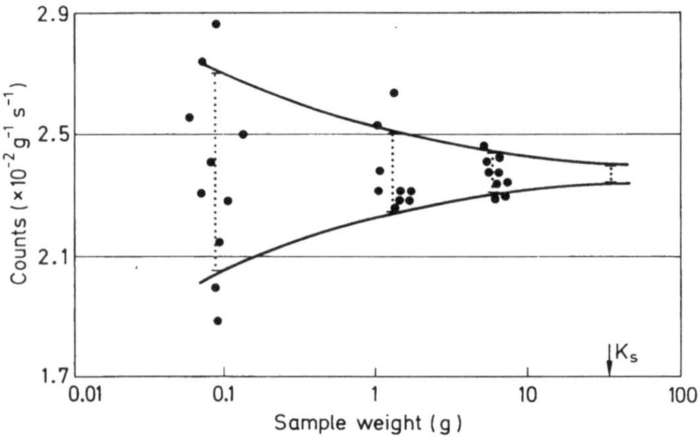

Fig. 4. Sampling diagram of sodium-24 in human liver homogenate [27]. Reprinted with permission from S. H. Harrison, R. Zeisler, NBS Internal Report 80-2164, p. 66, 1980. Copyright 1980 National Bureau of Standards

Gy [22] collected the prevailing techniques in his book on sampling particulate materials and developed an alternative way of calculating the sample size.

He defined a shape factor f as the ratio of the average volume of all particles having a maximum linear dimension equal to the mesh size of a screen to that of a cube which will just pass the same screen. f = 1.00 for cubes and 0.524 for spheres. For most materials f ~ 0.5. The particle size distribution factor g is the ratio of the upper size limit (95% pass screen) to the lower size limit (5% pass screen). For homogeneous particle sizes g = 1.00. The composition factor c is:

$$c = (1 - x)\,[(1 - x)\,d_x + xd_g]/x$$

where x is the overall concentration of the component of interest, d_x the density of this component and d_g the density of the matrix (remaining components). c is in the range $5 \cdot 10^{-5}$ kg/m^3 for high concentrations of c to 10^3 for trace concentrations.

The liberation factor l is

$$l = (d_1/d)^{1/2}$$

where d_1 is the mean diameter of the component of interest and d the diameter of the largest particles. The standard deviation of the sample s is estimated by

$$s^2 = fgcld^3/w$$

Ingamells [25] related Gy's sampling constant to Ingamells' constant by

$$K_s = fgcl(d^3 \times 10^4)$$

Brands proposed a calculation method in the case of segregation [23, 28]. A special type of inhomogeneous, particulate objects is the surface analysis by microscopic techniques e.g. analytical electron spectroscopy, laser induced mass spectroscopy or proton-induced X-ray emission. Here the minimum sample size can be translated into the minimum number of specific sample points in the specimen under investigation.

Morrison [30, 31] defined a sampling constant

$$K = (1/I_T)\left(A_T \sum_{i=1}^{n} i_i^2\right)^{1/2} (1 - A_I/A_T)^{-1/2}$$

where K = sampling constant in µm
i_i = intensity of inclusion i
n = total number of inclusions
A_T = total area image in µm^2
A_I = total area inclusions

For a confidence interval Δ the number of replicate analyses can be calculated

$$N = (100tK/\Delta^{1/2})^2$$

where a = area sampled
t = Students' t

Inczédy [32] proposed a measure for homogeneity of samples, assuming a sinusoidal change of concentrations of the component of interest.

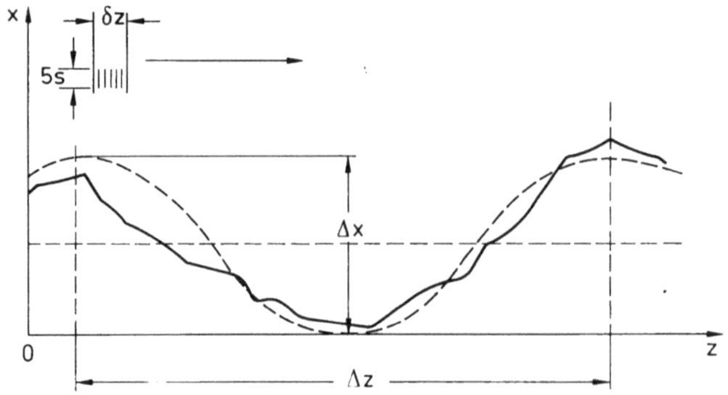

Fig. 5. Concentration distribution of an element in a solid. The resolution of the instrument is high compared to the concentration change. The error is also very low ($\Delta z \gg \delta z$; $\Delta x \gg s$). The vertical lines indicate the uncertainty of the measurement; the standard deviation s can be taken as a fifth of the range of the signals for a given level [32]. Reprinted with permission from Talanta, *29*. J. Inczédy, "Homogeneity of solids: a proposal for quantitative definition". Copyright 1982 Pergamon Press

The concentration difference Δc can be calculated by

$$\Delta c < 4.12s(\pi\,\delta z/\Delta z)/(\sin(\pi\,\delta z/\Delta z))$$

where s = measurement error
$\delta z/\Delta z$ = relative "window" (see Fig. 5)

Danzer et al. [33] proposed another homogeneity criterum based on the F test for samples of different size.

3.2 Internally Correlated Objects

Objects can be internally correlated in time or space. For mixed objects such as tanks (and the products that have been subjected to storage for longer or shorter times), rivers, lakes, gases and environmental air this correlation can be rendered by autocorrelagrams and modelled by a simple negative exponential function of the correlation distance.

The autocorrelation function can be calculated by:

$$\varrho_\tau = \left[\sum_{t=1}^{n-\tau} (x_t - \mu)(x_{t+\tau} - \mu)\right] \Big/ n \cdot \sigma^2$$

where x_t = measured value at time t
$x_{t+\tau}$ = measured value at time $t + \tau$
τ = time increment
n = number of measurements
σ^2 = variance of the measurements
ϱ_τ = 1 for $\tau = 0$ and approaches 0 for larger values of τ.

Assuming the exponential decay of ϱ_t the autocorrelation function can be represented by T_x, the time constant or correlation constant of the measured system as $\varrho_\tau = \exp(- /Tx)$, $Tx = \tau$ for $\varrho_\tau = 0$. See for instance [34].

Unmixed, stratified objects such as soil and rock, usually cannot be represented by a simple correlation constant derived from the exponential autocorrelation function. Here another measure is introduced, the semivariance. The semivariance γ_τ can be computed from [35-39].

$$\gamma_\tau = \sum_{t=1}^{n-\tau} (x_t - x_{t+\tau})^2/2n$$

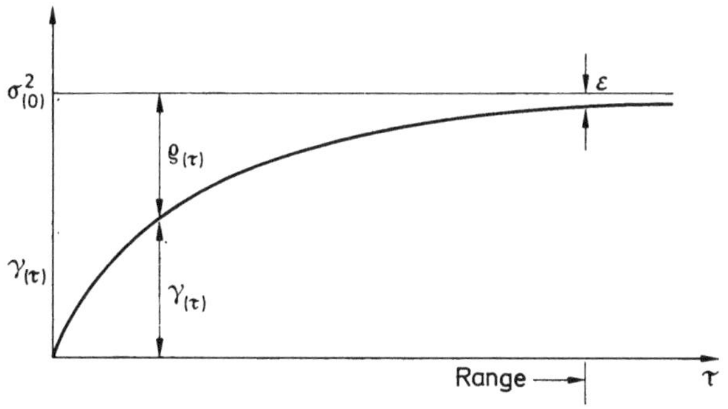

Fig. 6. Relationship between semivariance γ and autocovariance ϱ for a stationary regionalized variable. σ_0^2 is the variance of the observations, or the autocovariance at lag 0. For values of τ beyond the range, $\gamma(\tau) = \sigma_0^2$

The distance where γ_τ approaches σ^2 is called the range, a.

Fitting a model equation to an experimental semivariogram is a trial-and-error process, usually done by eye. Clark [40] describes and gives examples of the manual process, while Olea [41] provides a program which computes a model.

Some models used in practice are the linear model:

$$\gamma\tau = \alpha\tau \quad \text{for} \quad \tau < a$$

$$\gamma\tau = \sigma^2 \quad \text{for} \quad \tau \geqq a$$

The exponential model is given by

$$\gamma_\tau = \sigma^2(1 - \exp(-\tau/a))$$

and the "ideal" model is the spherical model

$$\gamma_\tau = \sigma^2(3\tau/2a - \tau^3/2a^3)$$

For objects that can be modelled by the time or correlation constant (see e.g. [42, 43]), Tx, the sample size, sample number and standard deviation are related and depend on the size of the object, if this object has a finite size in correlation units.

According to Müskens [44], the variance σ_*^2 can be thought of as composed of the variance in the composition of the sample σ_m^2, the variance in the composition of the whole object σ_μ^2 and the covariance between m and μ:

$$\sigma_*^2 = \sigma_m^2 + \sigma_\mu^2 - 2\sigma_{m\mu}$$

These variances can be calculated with

$$\sigma_m^2 = \frac{2\sigma_x^2}{ng}\left\{g - 1 + \exp(-g) + [\exp(-g) + \exp(g) - 2] \times \left(\frac{\exp(-a)}{1 - \exp(-a)} - \frac{\exp(-a)[1 - \exp(-p)]}{n(1 - \exp(-a))^2}\right)\right\}$$

$$\sigma_\mu^2 = \frac{2\sigma^2}{p^2}[p - 1 + \exp(-p)]$$

$$\sigma_{m\mu}^2 = \frac{\sigma_x^2}{npg}\left\{2ng + [1 - \exp(-p)]\left(\frac{\exp(-g) - 1}{1 - \exp(-a)} + \frac{\exp(g) - 1}{1 - \exp(a)}\right)\right\}$$

where σ_x^2 = variance of process
p = P/T_x
g = G/T_x
a = A/T_x
T_x = correlation factor of the process
n = number of samples

As can be seen in these equations, σ_*^2 depends on a number of factors. The properties of the population from which the object stems are described by σ_x and T_x. The relevant property of the object is its size p, expressed in units Tx. The relevant properties of the sample are increment size g, the number of increments n and the distance between the increments, a. If the sample size is expressed as a fraction of the object, F, the relations between F, σ_* and n are depicted in Fig. 7 and 8 [45].

The estimates of the sample size obtained in this way are valid for one-dimensional objects, e.g. output of factories, rivers, sampling lines in lakes, stationary sampling points for air monitoring, etc. A sample of papers that are devoted to the application of autocorrelation in sampling schemes is e.g. [46, 47, 48].

In soil science one usually has to do with two or threedimensional objects that cannot be represented by correlation constants. Here the size and the number of samples must be obtained in other ways. In this case often an intermediate step is Kriging, mapping of lines or planes of equal composition [41]. For the simplest case, punctual

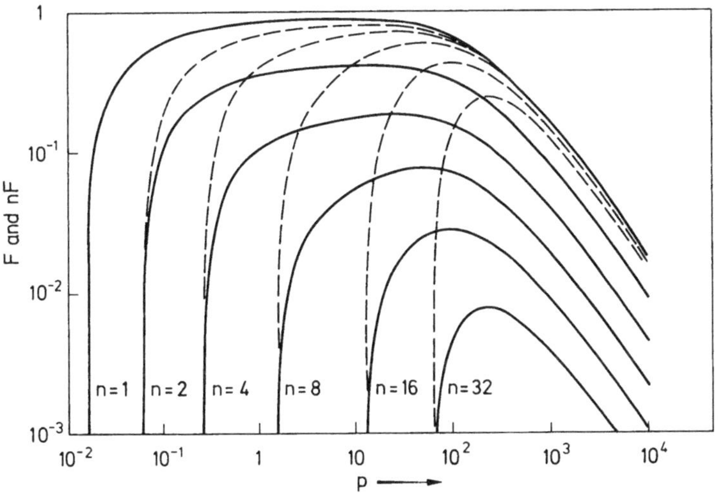

Fig. 7. The relative sample size F (————) and the relative gross sample size nF (------------) as a function of the relative lot size p for various number of samples (σ_*/σ_x) = 0.1 [45]. Reprinted with permission from G. Kateman, P. J. W. M. Müskens, Anal. Chim. Acta *103* (1978), p. 14. Copyright 1978 Elsevier Science Publishers B.V.

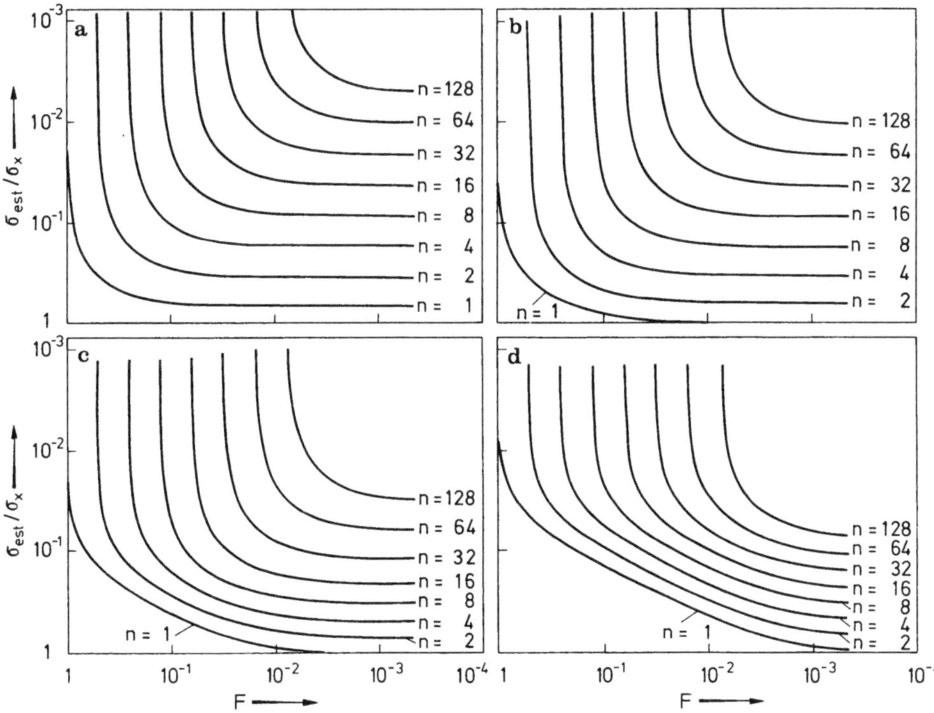

Fig. 8. Relationship between σ_*/σ_x and the sample size G for a given lot size (P) and time constant (T_x) and for different numbers of samples (n) [45]. (a) P = 1000, T_x = 1000, p = 1; (b) P = 1000, T_x = 100, p = 10; (c) P = 1000, T_x = 10, p = 100; (d) P = 1000, T_x = 1, p = 1000. Reprinted with permission from G. Kateman, P. J. W. M. Müskens, Anal. Chim. Acta *103* (1978), p. 16. Copyright 1978 Elsevier Science Publishers B.V.

Kriging, the composition of points p is estimated by solving W from the matrix equation

$$[A] \cdot [W] = [B]$$

$$A = \begin{bmatrix} \gamma(\tau 11) & \gamma(\tau 12) & \ldots & 1 \\ \gamma(\tau 12) & \gamma(\tau 22) & \ldots & 1 \\ \gamma(\tau 13) & \gamma(\tau 23) & \ldots & 1 \\ 1 & 1 & 1 & 0 \end{bmatrix}$$

with $\gamma(\tau ij)$ the semivariance over a distance τ corresponding to the control points i and j.

$$W = \begin{bmatrix} W1 \\ W2 \\ \vdots \\ \lambda \end{bmatrix} \qquad B = \begin{bmatrix} \gamma(\tau 1p) \\ \gamma(\tau 2p) \\ \vdots \\ 1 \end{bmatrix}$$

with $\gamma(\tau 1p)$ the semivariance over a distance τ equal to that between known point i and the location p where the estimate is to be made.

Now $x_p = W_1x_1 + W_2x_2 + W_3x_3$ with a variance
$\sigma_\varepsilon^2 = W_1\gamma(\tau 1p) + W_2\gamma(\tau 2p) + W_3\gamma(\tau 3p) + \lambda$

The map that can be obtained in this way shows not only the estimated value at some interpolated point p but also its error. With this map it can be decided where new samples should be taken.

By estimating "compartments" of compositions that do not vary more than a given amount, sampling can be restricted to one sample per compartment. Most compartment estimates are arbitrary, however [49, 50, 51, 52].

4 Sampling for Detailed Description and Control

When the objective of sampling is estimation of the composition of the object in detail, the sampling strategy will be different from the strategy aimed at the estimation of the mean composition of the lot. Shannons sampling theorem states that a signal (in analytical chemistry an object) should be sampled with a frequency of 2 times the highest frequency present in the signal. Only in that case a reliable estimate of the real situation can be obtained. The highest frequency can be deducted from a Fourier transform of the signal.

For analytical chemical purposes this approach is not very practical. To obtain sufficient data for a reliable Fourier analysis can be a hard task.

Estimating the autocorrelogram is easier in most cases and use of a model of the distribution of the component allows reconstruction of the value of the composition.

Application of these reconstruction methods [53, 54] can be described as the interpolation of object composition between sample points by means of an exponential function, characterised by the correlation constant.

As Van der Grinten developed his estimation algorithm for real time process control he also introduced a "dead time", the time between sampling and availability of the analytical result. As a measure of reconstruction efficiency he defined the measurability factor

$$m = ((\sigma_x^2 - \sigma_e^2)/\sigma_x^2)^{1/2}$$

where σ_x^2 = variance of composition of object
σ_e^2 = residual variance after reconstruction

When the object is a process, an object whose composition changes in time, this factor m can be estimated by:

$$m = [\exp - (d + 1/2a + 1/3y)]\,(1 - sa \cdot t_e^{1/2})$$

where $d = D/T_x$ $a = A/T_x$ $g = G/T_x$ $t_e = T_e/T_x$ and $s = \sigma_a/\sigma_x$
D = analysis time
A = (sampling frequency)$^{-1}$
G = sample size
T_x = time constant (correlation factor) of process
T_e = time constant of measuring device
σ_a^2 = variance of method of analysis
σ_x^2 = variance of process

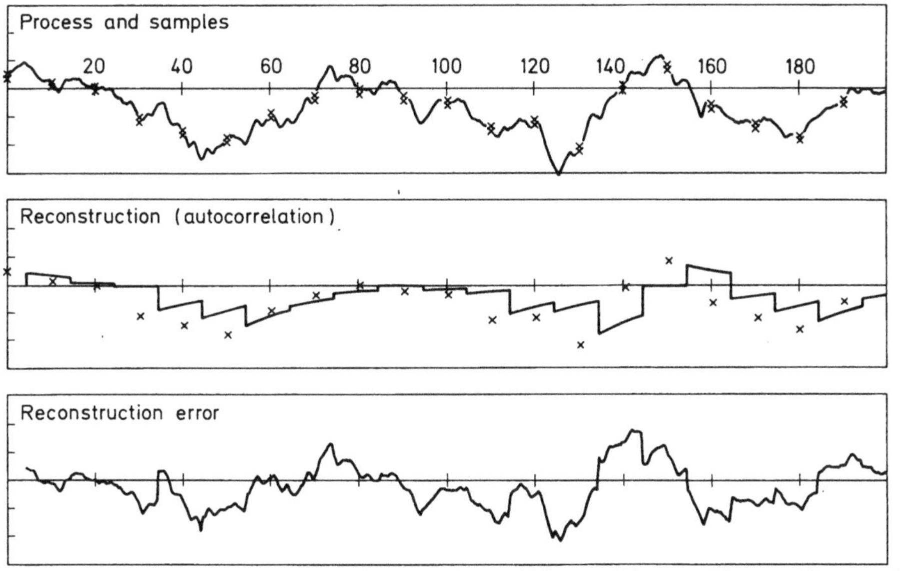

Fig. 9. Reconstruction and reconstruction error

The equation can be rewritten as:

$$m = m_D \cdot m_A \cdot m_G \cdot m_N$$

where $m_D = \exp(-d)$
$m_A = \exp(-a/2)$
$m_G = \exp(-g/3)$
$m_N = 1 - s_a t_e^{1/2}$

From these equations it follows that the maximum obtainable measurability factor will never exceed the smallest of the composing factors. This implies that all factors should be considered in order to eliminate the restricting one. It also means that a trade-off is possible between high and low values of the various factors.

Leemans [55] described a sampling scheme based on these algorithms that considers sampling frequency, sampling time, dead time and accuracy of the method of analysis to obtain optimal information yield or maximal profit when controlling å factory.

The sampling rate and therefore the information yield is not only set by the sampling scheme, but also by the laboratory. If, for instance, the frequency of sampling is too high queueing of samples occur, that means loss of information if the results are used for process control.

Janse [56] showed that the theory of queueing can be applied to study the effects on information yield of such limited facilities. The effect depends on the way the sampling and analysis are organized.

The effect can be calculated for simple systems with one service point (one analyst or one instrument). For more complicated systems simulation should be applied. For a M/M/1 system (random sampling/random analysis time/1 server), in fact the most unorganized system, the measurability factor is:

$$m^2 = [T_x/(T_x + 2\bar{A})]\ \{(1 - \varrho)\, T_x/[(1 - \varrho)\, T_x + \bar{D}]\}$$

where T_x = time constant (correlation constant)
$\bar{A}$ = mean interarrival time of samples
$\bar{D}$ = mean interanalysis time
ϱ utilization factor of the lab

For a D/D/1 system (fixed sampling rate/fixed analysis time) there are no waiting times.
The situation D/M/1 (fixed sampling rate) and M/D/1 (fixed analysis time) are intermediate.

D/M/1 system $m^2 = \exp(-2\bar{A}/T_x)\ \{(1 - \varrho)\, T_x/[(1 - \varrho)\, T_x + 2D]\}$

M/D/1 system $m^2 = [T_x/(T_x + 2\bar{A})]$
$\{2(1 - \varrho)\, \bar{A}/[(2\bar{A} - T_x) \exp(2\bar{D}/T_x) + T_x]\}$

One obvious conclusion can be that smoothing the input (from M/M/1 to D/M/1) is most advantageous.

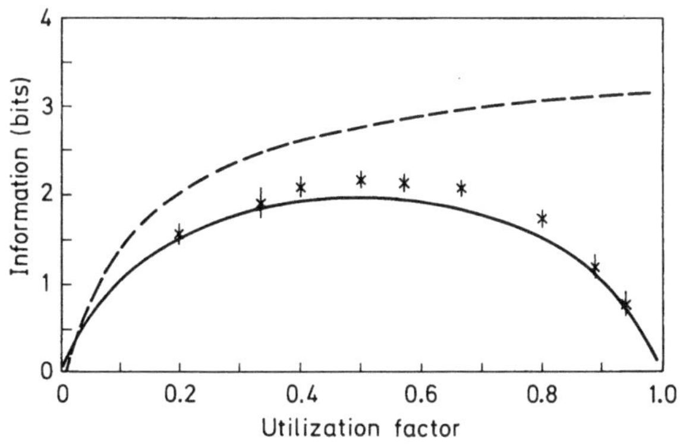

Fig. 10. Theoretical curves for information yield as a function of utilization factor: (————) M/M/1 system; (·············) D/D/1 system; (×) simulation results with 95% probability intervals [56]. Reprinted with permission from T. A. H. M. Janse, G. Kateman, Anal. Chim. Acta *150*, p. 225. Copyright 1983 Elsevier Science Publishers B.V.

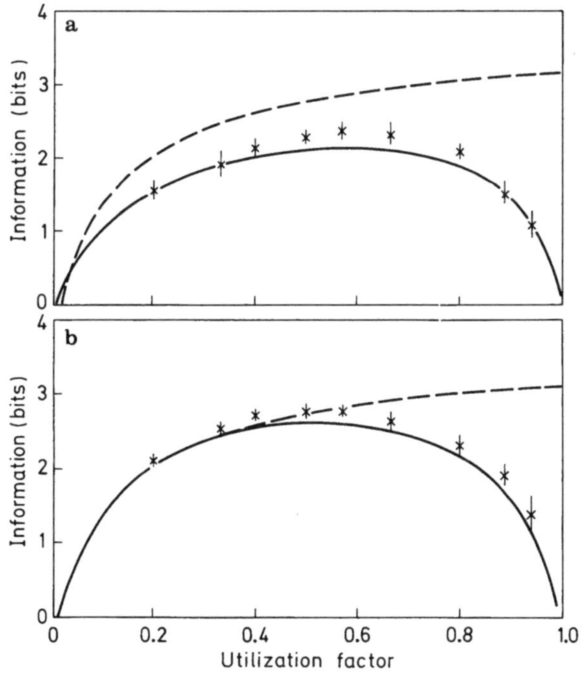

Fig. 11. Theoretical curves for information yield as a function of utilization factor: A, M/D/1 system; B, D/M/1 system. (·············) D/D/1 system; (×) simulation results with 95% probability intervals [56]. Reprinted with permission from T. A. H. M. Janse, G. Kateman, Anal. Chim. Acta *150*, p. 226. Copyright 1983 Elsevier Science Publishers B.V.

5 Sampling for Monitoring

If the objective of sampling is to provide information for warnings (threshold monitoring) autocorrelated processes can be modelled by the earlier described methods. Not only interpolation is possible but extrapolation can be applied as well. However, the uncertainty in the extrapolated estimate depends on the prediction time. As long as the predicted value, including the prediction error does not exceed the preset warning threshold, no new sample is required. Müskens [57] derived that the next sample should be taken at a time τ after an analytical result x_t according to:

$$\tau = T_x \cdot \ln \left\{ \frac{T_r \cdot X_t + Z[X_t^2 - q(T_r^2 - Z^2)]^{1/2}}{X_t^2 + qZ^2} \right\}$$

where T_r = threshold value (normalized to zero mean and unit standard deviation)
X_t = process value (normalized to zero mean and unit standard deviation)
Z = reliability factor
$q = (\sigma_x^2 + \sigma_a^2)/\sigma_x^2$
σ_x^2 = variance process
σ_a^2 = variance method of analysis

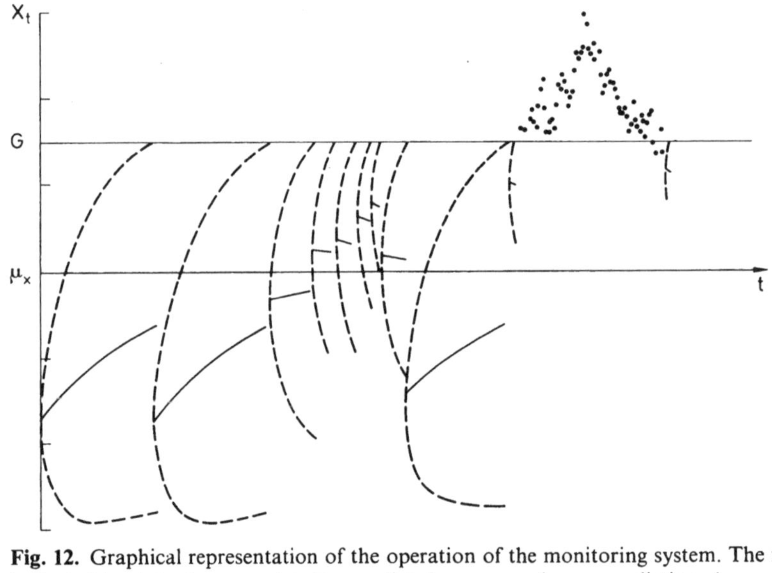

Fig. 12. Graphical representation of the operation of the monitoring system. The measurements are indicated by dots. (————) is the predicted process value or prediction. (------------) indicates the reliability interval of the prediction. Here the 95% interval is used. Reprinted with permission from P. J. W. M. Müskens, Anal. Chim. Acta *103* (1978), p. 447. Copyright 1978 Elsevier Science Publishers B.V.

He also derived mean sampling rates when a known probability of exceeding the threshold without sampling is accepted.

6 Other Objectives

When the objective is not solely maximizing the information from samples but when economic or organizational aspects are at stake as well, there are more ways to influence sampling schemes. Next to the influences of sampling rate as already described there are possibilities in collecting samples in batches and assigning priorities to samples. The influence on laboratory performance and information content has been treated by Vandeginste and Janse [58,59,60].

7 Conclusions

The remarkable growth of chemometrics in the past 15 years had its influence on sampling strategies. The older techniques relied on the static statistical properties of the object under observation. The newer techniques all take into account the correlations within the object, allowing optimal sampling of processes. No doubt these techniques have been made possible by the availability of computers. Developments now seem to concentrate on refinements of existing techniques, allowing the optimal design of sampling networks, e.g. for environmental monitoring of air and water or soil survey for agriculture and geology.

8 References

1. Arbeitskreis "Automation in der Analyse": Z. Anal. Chem. *261*, 1 (1972)
2. Kratochvil, B., Wallace, D., Taylor, J. K.: Anal. Chem. *56*, 113R (1984)
3. Ku, H. H.: NBS Spec. Pub. *519*, 1 (1979)
4. ACS Committee on Environmental Improvement: Anal. Chem. *55*, 2210 (1983)
5. Kateman, G., Pijpers, F. W.: Quality Control in Chemical Analysis, Wiley, New York 1981
6. Kateman, G.: Sampling, in: Chemometrics, mathematics and statistics in chemistry (Kowalski, B. R., ed.) p. 177, Reidel Dordrecht 1984
7. Cochran, W. B.: Sampling Techniques, Wiley, New York 1977
8. Kratochvil, B., Taylor, J. K.: Anal. Chem. *53* (8), 924A (1981)
9. Smith, R., James, G. V.: The sampling of bulk materials, Royal Soc. of Chem., London 1981
10. Bicking, C. A., in: Treatise on analytical chemistry (Kolthoff, I. M., Elving, P. J., ed.) p. 299, Wiley, New York 1979[2]
11. Horwitz, W. J.: J. Assoc. Off. Anal. Chem. *59*, 238 (1976)
12. Pijper, J. W.: Anal. Chim. Acta *170*, 159 (1985)
13. Illingworth, F. K.: Trends in Anal. Chem. *4* (5), IX (1985)
14. Kratochvil, B. G., Taylor, J. K.: NBS Tech. Note 1153, Nat. Bureau of Standards, Washington DC 1982
15. Bauer, C. F.: J. Chem. Educ. *62*, 253 (1985)
16. Kratochvil, B., Reid, R. S.: ibid. *62*, 252 (1985)
17. Kateman, G., Van der Wiel, P. F. A., Janse, T. A. H. M., Vandeginste, B. G. M.: CLEOPATRA, Elsevier Scientific Software, Amsterdam 1985
18. Baule, B., Benedetti-Pichler, A. A.: Z. Anal. Chem. *74*, 442 (1928)
19. Visman, J.: Mat. Res. Stds. *9* (11), 8 (1969)
20. Visman, J., Duncan, A. J., Lerner, M.: ibid. *11* (8), 32 (1971)
21. Visman, J.: J. Mat. *7*, 345 (1972)

22. Gy, M.: Sampling of particulate materials: theory and practice, Elsevier, Amsterdam 1979
23. Harris, W. E., Kratochvil, B.: Anal. Chem. *46*, 313 (1974)
24. Ingamells, C. O., Switzer, P.: Talanta *20*, 547 (1973)
25. Ingamells, C. O.: ibid. *21*, 141 (1974)
26. Ingamells, C. O.: ibid. *23*, 263 (1976)
27. Harrison, S. H., Zeisler, R.: "NBS Internal report 80-2164", p. 66, US National Bureau of Standards, Washington 1980
28. Brands, G.: Fres. Z. Anal. Chem. *314*, 6 (1983)
29. Brands, G.: ibid. *314*, 646 (1983)
30. Fasset, J. D., Roth, J. R., Morrison, G. H.: Anal. Chem. *49*, 2322 (1977)
31. Scilla, G. J., Morrison, G. H.: ibid. *49*, 1529 (1977)
32. Inczédy, J.: Talanta *29*, 643 (1982)
33. Danzer, K., Doerffel, K., Ehrhardt, H., Grissler, M., Ehrlich, G., Gadow, P.: Anal. Chim. Acta *105*, 1 (1979)
34. Box, G. E. P., Jenkins, G. M.: "Time Series Analysis", Holden-Day, San Francisco 1970
35. Armstrong, M., Jabin, R.: Math. Ged. *13*, 455 (1981)
36. David, M.: "Geological Ore Reserve Estimation", Elsevier, Amsterdam 1977
37. Journel, A. G., Huijbregts, C. J.: "Mining Geostatistics", Academic Press London 1978
38. Davis, J. C., McCullagh, M.: "Display and Analysis of Spatial Data", Wiley, New York 1975
39. Davis, J. C., in: (Kowalski, B. R., ed.) "Chemometrics, mathematics and statistics in chemistry", p. 419, Reidel Dordrecht 1984
40. Clark, I.: "Practical Geostatistics", Applied Science Publishers, Ltd., London 1979
41. Olea, R. A.: "Measuring spatial dependence with semivariograms", "Kansas Geological Survey Series on Spatial Analysis", no. 3, University Kansas, Lawrence
42. Vandeginste, B. G. M., Salemink, P. J. M., Duinker, C. J.: Neth. J. Sea Res. *10*, 59 (1976)
43. Müskens, P. J. W. M., Hensgens, W. G. J.: Water Res. *11*, 509 (1977)
44. Müskens, P. J. W. M., Kateman, G.: Anal. Chim. Acta *103*, 1 (1978)
45. Kateman, G., Müskens, P. J. W. M.: ibid. *103*, 11 (1978)
46. Bobec, B., Bluis, D.: J. Hydrol. *44*, 17 (1979)
47. Bobec, B., Lachance, M., Cazaillet, O.: Eau Que *16*, 39 (1973)
48. Nelson, J. D., Ward, R. C.: Groundwater *19*, 617 (1981)
49. Bouma, J., in: "Soil Spatial Variability" (Nelson, D. R., Bouma, J., ed.) p. 130, Wageningen 1985
50. Gruijter, J. J. de, Marsman, B. A., in: "Soil Spatial Variability" (Nelson, D. R., Bouma, J., ed.) p. 150, Wageningen 1985
51. Bouma, J.: Agric. Water Managm. *6*, 177 (1983)
52. Webster, R., Burgess, T. M.: ibid. *6*, 111 (1983)
53. Grinten, P. M. E. M. van der: J. Instr. Soc. Am. *12* (1), 48 (1965)
54. Grinten, P. M. E. M. van der: ibid. *13* (2), 58 (1966)
55. Leemans, F. A.: Anal. Chem. *43* (11), 36A (1971)
56. Janse, T. A. H. M., Kateman, G.: Anal. Chim. Acta *150*, 219 (1983)
57. Müskens, P. J. W. M.: ibid. *103*, 445 (1978)
58. Vandeginste, B. G. M.: ibid. *122*, 435 (1980)
59. Vollenbroek, J. G., Vandeginste, B. G. M.: ibid. *133*, 85 (1981)
60. Janse, T. A. H. M., Kateman, G.: ibid. *159*, 181 (1984)

Signal and Data Analysis in Chromatography

Henri Casemirus Smit and Erik Jan van den Heuvel

Laboratory for Analytical Chemistry, University of Amsterdam, Nieuwe Achtergracht 166, 1018 WV Amsterdam, The Netherlands

Table of Contents

An overview is given of chromatographic signal and data processing techniques, including noise analysis and uncertainty calculations. In the single channel signal processing, attention is paid to the use of signal approximation and regression with different signal models. In the multivariate approach recently developed deconvolution methods are described, applying a.o. least squares optimization and factor analysis. Finally, chemometric optimization and classification procedures are treated, using chromatographic data sets and special techniques like correlation chromatography.

1 Introduction

The chromatographic process involves the physical separation of substances, a measuring procedure (detection), followed by signal — as well as data processing. Column chromatography ideally yields a number of well-separated peaks with an approximately Gaussian shape and a flat noiseless baseline. The ideal situation is pursued by adjusting and optimizing a set of in general mutual-dependent parameters like flow, temperature, concentration.

The relevant quantitative or qualitative information is extracted from the obtained signal by, for example, determining peak height or peak area and retention time. The simple manual data processing procedures of former days are now automized and computerized and data acquisition involving digitizing the originally analog detector signal is incorporated.

In practice, the real signal is never ideal. Systematic and random errors often occur due to, e.g. unresolved or badly resolved peaks, non-linearity (resulting in concentration-dependent peak shapes), noise and drift.

The mentioned simple data processing, suitable for perfect or almost perfect chromatograms, is not optimal in case of "difficult" chromatograms. More advanced data and signal processing procedures are developed, resulting in lower systematic and random errors.

The quality of the chromatographic procedure or parts of it can be optimized, for instance by using a multi-wavelength detector or a mass spectrometer.

Advanced dataprocessing and multiple signal detectors require a chemometrical approach. However, the use of chemometrics is not restricted to these applications. The chromatographic parameters can be optimized by using statistical and mathematical techniques like factor analysis and formal optimization techniques. And even the normal chromatographic process can be modified chemometrically by, for instance, replacing the single injection by a random injection pattern and applying correlation techniques (correlation chromatography). The result is a drastically decreased detection limit.

Further, analytical dataprocessing has to be completed with an estimation of the systematic error and the uncertainty in the results. Signal and noise analysis, required for this purpose, is based on mathematical and statistical techniques and can therefore be classified as a part of chemometrics. Moreover, determining signal and noise characteristics by developing mathematical (statistical) models is indispensable in dataprocessing procedures; advanced dataprocessing is in general based on preknowledge of signal and noise.

2 Chromatographic Signal Analysis

A chromatogram without noise and drift is composed of a number of approximately bell-shaped peaks, resolved and unresolved. It is obvious that the most realistic model of a single peak shape or even the complete chromatogram could be obtained by the solution of mass transport models, based on conservation laws. However, the often used plug flow with constant flow velocity and axial diffusion, resulting in real Gaussian peak shape, is hardly realistic. Even a slightly more complicated transport equation with, for instance, a non-linear partition isotherm, cannot be solved mathematically. The only possibility is a computer simulation.

Both a simplified continuous and discrete model, describing the behaviour of single component mass transport in chromatographic columns with non-linear distribution isotherm, were developed and simulated by Smit et al. [1,2,3]. Studies of more complex but still relatively simple (multicomponent) transport models have been published (see e.g. [4,5]).

The alternative is the use of a descriptive mathematical model without any relation with the solution of the transport equation. On the analogy of the characterization of statistical probability density functions a peak shape f(t) can be characterized by *moments*, defined by:

$$m_k = E[t^k] = \int_{-\infty}^{+\infty} t^k f(t)\, dt \qquad k = 0, 1, 2 \ldots \tag{1}$$

E[ ] denotes the expected value of the expression between the brackets [6]. The constants:

$$\mu_k = E[(t - m_1)^k] \tag{2}$$

are called the central moments and can be expressed in $\{m_j\}_j^k$ by expanding Eq. (2) by the binomial theorem:

$$\mu_k = E[(t - m_1)^k] = E\left[\sum_{r=0}^{k} \binom{k}{r} (-1)^r\, m_1^r t^{k-r}\right] \tag{3}$$

Particularly important are m_0 the peak area, m_1 the centre of gravity, and $\mu_2 = \sigma^2$, where σ is the standard deviation of the peak.

Different peak shapes can also be compared by the determination of dimensionless moments about the mean, defined as:

$$a_r = \frac{\mu_2}{\sqrt{\mu_2^r}} \tag{4}$$

Particularly important are the moment coefficient of skewness:

$$a_3 = \frac{\mu_3}{\sqrt{\mu_2^3}} \tag{5}$$

and the moment of kurtosis, the degree of peakedness:

$$a_4 = \frac{\mu_4}{\mu_2^2} \tag{6}$$

A related approach is the approximation of peak-shaped functions by means of orthogonal polynomials, described by Scheeren et al. [7]. A function f(t), in this case the chromatographic signal, can be expanded in a series:

$$f(t) = a_0 p_0(t) + a_1 p_1(t) + \ldots + a_n p_N(t) \tag{7}$$

Multiplying both sides with $p_n(t)$ and integrating gives:

$$\int_{-\infty}^{\infty} f(t)\, p_n(t)\, dt = a_0 \int_{-\infty}^{\infty} p_0(t)\, p_n(t)\, dt + \ldots + a_n \int_{-\infty}^{\infty} p_N(t)\, p_n(t)\, dt \qquad (8)$$

Suppose f(t) is known from experimental data, then for each function $p_n(t)$ the left-hand side can be calculated. This procedure can be repeated N + 1 times, giving N + 1 results, usable to calculate the coefficients a_0, a_1, ... a_N. Important criteria for the choice of the function $p_n(t)$ are the fast convergence and the accuracy of the method.

If the functions 1, t, t^2, ... t^N are chosen, then the already mentioned moments of f(t) are found. However, convergence is not guaranteed in this case. Moreover, the calculation of the coefficients a_n requires the solution of N equations with N unknowns and the values a_n are dependent of N. The introduction of a specialized set of orthogonal polynomials can be advantageous and circumvents some problems. Suppose that the following integral exists:

$$\int_a^b |p_i(t)|^2\, w(t)\, dt < \infty \qquad i = n, m \qquad (9)$$

The scalar product of the functions $p_n(t)$ and $p_m(t)$ with respect to the weighting function w(t) is:

$$\langle p_n(t), p_m(t)\rangle = \int_a^b p_n(t)\, p_m(t)\, w(t)\, dt \qquad (10)$$

If the set $[p_i]$ satisfies the condition that $\langle p_n(t), p_m(t)\rangle = 0$ for $n \neq m$ then the series is called orthogonal, and orthonormal when in addition $\langle p_n(t), p_m(t)\rangle = 1$, for $n = m$.

Approximating a function with an orthogonal polynomial series means that it is not necessary to solve N equations simultaneously.

The addition of more terms does not influence the values of the already calculated terms. In this aspect, orthogonal polynomials are superior to other polynomials; calculation of the coefficients is simple and fast. Moreover, according to the Gram-Schmidt theory every function can be expressed as a series of orthogonal polynomials, using the weighting function w(t).

The choice of the specific orthogonal polynomial is determined by the convergence. If the signal to be approximated is a bell-shaped function, it is evident to use a polynomial derived from the Gauss function, i.e. one of the so-called classical polynomials, the Hermite polynomial. Widely used is the Chebyschev polynomial; one of the special features of this polynomial is that the error will be spread evenly over the whole interval.

Figure 1 shows an example of a chromatogram of alkylbenzenes approximated with Chebyschev series. The information present in the chromatogram is reflected in the values of the polynomial coefficients. However, the terms do not usually have a direct relationship with conventional analytical parameters.

A typical application is given by Debets et al. [8]. A quality criterion for the characterization of separation in a chromatogram is modified by using Hermite polynomial coefficients in order to enhance the performance. The quality criterion can be used in

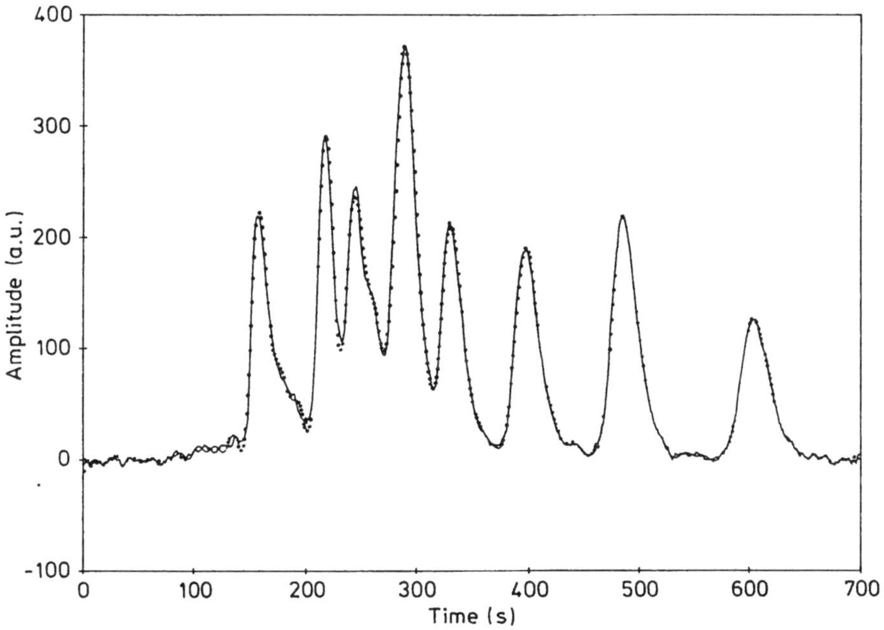

Fig. 1. Chromatogram of alkylbenzenes (reversed-phase HPLC) 100 term Chebyshev polynomial approximation.

an automatic optimization procedure in HPLC. Figure 2 (a and b) shows the effect of the addition of a coefficient on the change in the quality of separation with strongly overlapping peaks.

Related to the approximation of signals with orthogonal series is the widely used description in the frequency domain. Given a function of the time t, one can form the integral:

$$F(\omega) = \int_{-\infty}^{\infty} f(t)\, e^{-j\omega t}\, dt \tag{11}$$

If the integral exists, Eq. (11) defines a generally complex function $F(\omega)$, known as the Fourier transform of $f(t)$:

$$F(\omega) = R(\omega) + jX(\omega) = A(\omega)\, e^{j\varphi(\omega)} \tag{12}$$

$A(\omega)$ is called the Fourier spectrum, $A^2(\omega)$ is the energy spectrum and $\varphi(\omega)$ the phase angle of $f(t)$.

Another approach is the characterization of peaks with a well-defined model with limited parameters. Many models are proposed, some representative examples will be described. Well known is the Exponentially Modified Gaussian (EMG) peak, i.e. a Gaussian convoluted with an exponential decay function. Already a few decades ago it was recognized that an instrumental contribution such as an amplifier acting as a first-order low pass system with a time constant, will exponentially modify the

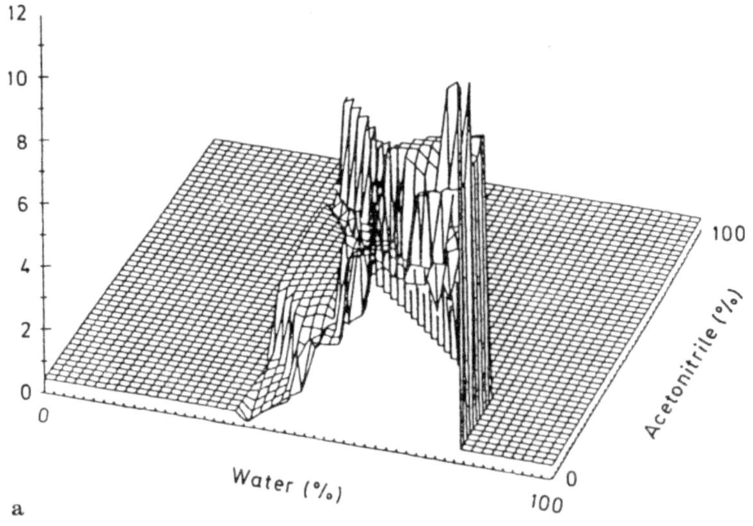

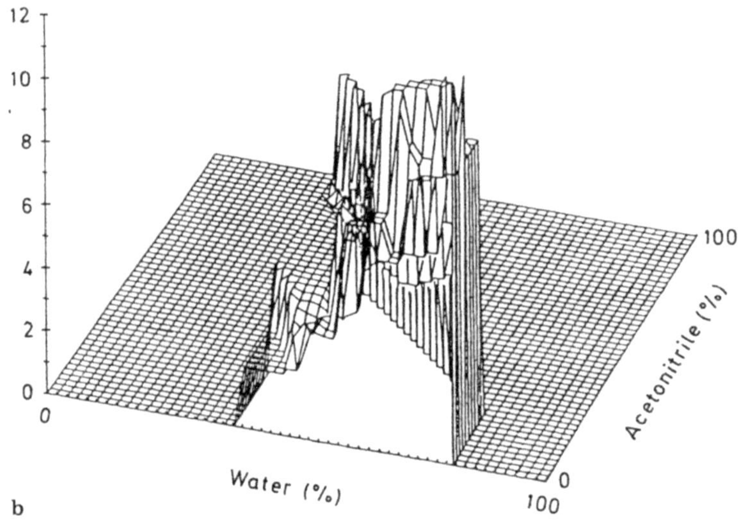

Fig. 2. Response surfaces of a separation quality criterion from chromatograms of sulfanilamide, sulfacetamide, sulfadiazine, sulfisomidine and sulfathiazole, with eluents consisting of water, methanol and acetonitrile. (a) with, and (b) without Hermite polynomial coefficients.

chromatographic peak. Several authors described the EMG, most significantly this was done by Grushka [9] and Littlewood [10]. The EMG is defined by the following convolution:

$$f(t) = \frac{A}{\tau\sigma\sqrt{2\pi}} \int_{0}^{\infty} \exp\left[-\frac{(t - t_R - t')^2}{2\sigma^2}\right] \exp\left[-\frac{t'}{\tau}\right] dt' \qquad (13)$$

where
σ = standard deviation of the Gaussian
A = peak amplitude
τ = time constant of the exponential modifier
t_R = centre of gravity (top) of the Gaussian
t' = dummy integration variable
Grushka derived the moments of the EMG. The first moment is:

$$m_1 = t_R + \tau$$

The second, third and fourth central moments are:

$$\mu_2 = \sigma^2 + \tau^2$$

$$\mu_3 = 2\tau^3$$

$$\mu_4 = 3\sigma^4 + 6\sigma^2\tau^2 + 9\tau^4$$

The value of τ determines the asymmetry of the peak.

Yau [11] derived characteristical properties of the EMG skewed peak model and proposed a new method to extract band broadening parameters from a skewed and noisy peak.

Another model is suggested by Fraser and Suzuki [12]:

$$f(t) = H \exp\left[-\frac{\ln 2}{A^2}\left\{\ln\left(1 + \frac{A(t - t_R)}{\sigma(2\ln 2)^{1/2}}\right)\right\}^2\right] \tag{14}$$

where
H = peak height
A = asymmetry factor
σ = standard deviation
t_R = time at the peak maximum

The model gives a good approximation of practical signals. Figure 3 shows some examples. Also a model related to the Γ-distribution has been proposed [13]:

$$f(t) = K\left(\frac{t - t_0}{\tau}\right)^{(n-1)} e^{-\frac{t-t_0}{\tau}} \tag{15}$$

where K determines the peak amplitude, τ the peak width, n the asymmetry and t_0 the peak start. The four parameters of this function can be determined by a simple fitting procedure, even without a computer, and, moreover, an estimation of the moments of a chromatographic peak can be calculated from the parameters. The kth moment of the (normalised) gamma density function is:

$$m_k = \tau^k \frac{(k + n - 1)!}{(n - 1)!} \tag{16}$$

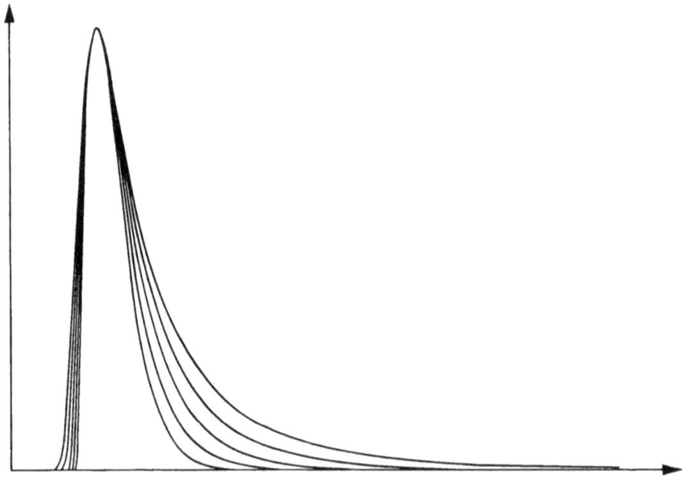

Fig. 3. Frazer-Suzuki models with different asymmetry factors.

The model is directly related to the widely used theoretical plate concept, which, in principle, is only valid for Gaussian peaks.

However, the Γ-distribution permits an extension of the plate theory, which is also usable in case of asymmetric peaks. The chromatogram (1 component) is considered to be the result of a pure time delay and a Γ-distribution response. The procedure implies the fitting of a function f(t) given in Eq. (15) to the chromatographic peak. The asymmetry of the peak determines the "new plate number" n′, decreasing with increasing asymmetry.

A transfer function, defined as the Laplace transfer of the impulse response of a linear system, can be obtained from the model. This can be very useful, because with a transfer function the influence of extra-column effects (detector, amplifier, filter) on the peak shape can be easily calculated. The transfer function is:

$$H(s) = K \cdot e^{-st_0} \frac{\Gamma(n')}{\tau^{n'-1}} \frac{1}{\left(s + \frac{1}{\tau}\right)^{n}} \tag{17}$$

where n′ is the new plate number, τ and K are constants, and t_0 is the time delay.

Addition of an amplifier (first-order system) with time constant τ_v and amplification K_2 gives

$$H(s) = K_1\, e^{-sT_d} \frac{\Gamma(n')}{\tau^{n'-1}} \frac{1}{\left(s + \frac{1}{\tau}\right)^{n'}} K_2 \frac{1/\tau_v}{\left(s + \frac{1}{\tau_v}\right)} \tag{18}$$

The Laplace inverse transform h(t) of Eq. (18) gives the shape of the impulse response, in this case an exponentially modified asymmetric chromatographic peak. The pth moment of h(t) can be determined by:

$$m_p = (-1)^p \frac{d^pF(o)}{ds^p}, \qquad \text{assuming } m_0 = 0 \tag{19}$$

From the moments m_p the central moments μ_p can be calculated. Some results, important in chromatography, are

$$\begin{aligned} m_1 &= \tau_v + n'\tau \\ \mu_2 &= m_2^2 - m_1^2 = \tau_v^2 + n'\tau^2 \\ \mu_3 &= m_3 - 3m_2m_1 + 2m_1^3 = 2(n'\tau^3 + \tau_v^3) \end{aligned} \tag{20}$$

Vaidya and Hester [14] have proposed a model, based on the modified Generalized Exponential (GEX) function:

$$F(t) = \frac{ac^{b/a}}{\Gamma(b/a)} e^{-ct^a} at^{(b-1)} \tag{21}$$

where a, b, and c are constants.

The function is very general in nature. Each peak can be represented by five parameters, one more than for instance the Γ-distribution.

3 Chromatographic Noise Analysis

In chromatography the quantitative or qualitative information has to be extracted from the peak-shaped signal, generally superimposed on a background contaminated with noise. Many, mostly semi-empirical, methods have been developed for relevant information extraction and for reduction of the influence of noise. Both for this purpose and for a quantification of the random error it is necessary to characterize the noise, applying theory, random time functions and stochastic processes. Four main types of statistical functions are used to describe the basic properties of random data:

— the mean and the mean square value, given by

$$\mu_x = \lim_{T\to\infty} \int_0^T x(t)\,dt \qquad \text{and} \qquad \psi_x^2 = \lim_{T\to\infty} \frac{1}{T} \int_0^T x^2(t)\,dt \tag{22}$$

x(t) = random function

— probability density functions (PDF)

— autocorrelation functions (ACF)

— power spectral densities (PSD)

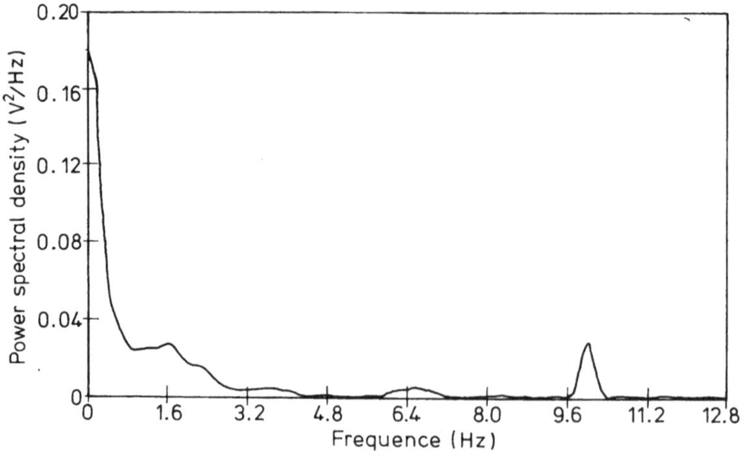

Fig. 4. Power spectrum of the noise of a FID.

Many authors are assuming white noise, which is hardly realistic in analytical practice and certainly not in chromatography. Noise can be classified in two classes: stationary and non-stationary. The statistical parameters of stationary noise do not vary with the time, this in contrast with non-stationary noise where one or more parameters, for instance the mean or the variance, are time-dependent.

In chromatography the noise is generally composed of stationary noise superimposed on a slowly varying baseline (drift). The slow baseline variation may be stochastic or deterministic. In general, proportional noise, i.e. noise with a standard deviation approximately proportional to the signal amplitude, can be neglected. It is noteworthy that sometimes the chromatographic noise, even including "drift", may be considered as approximately stationary, if observed during a long time. However, contamination of the detector, stripping of the column, etc. are often the origin of an irreversible drift component. At short term, during a chromatogram or a part of a chromatogram, the low frequency components of stationary noise are often considered as non-stationary drift.

Scheeren et al. [15] described the drift component of a chromatogram with a higher order polynomial. Some chromatographic noise analysis is done by Smit et al. [16], and later by Duursma et al. [17] and Laeven et al. [18]. One may conclude that chromatographic noise in general has a significant 1/f character (flicker noise), i.e. the noise power is approximately reversed proportional with the frequency (Fig. 4). Of course, real 1/f noise is impossible, because the spectrum shows a singularity at f = 0. A realistic noise model, suitable for describing many analytical noise sources is given by Walg and Smit [19]:

$$G(\omega) = c \cdot \frac{\arctan \omega t_n - \arctan \omega t_m}{\omega} \tag{23}$$

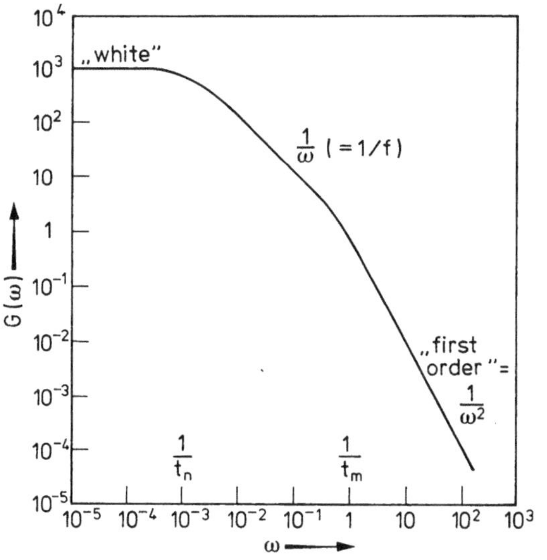

Fig. 5. Power Spectral Density of eq. (23) with $t_n = 10^3$ and $t_m = 1$.

where

ω = angular frequency

G(ω) = power spectral density

t_n, t_m = time constants

c = constant

Figure 5 shows the PSD with a constant "white" region, a $1/\omega$ (= 1/f) region and a $1/\omega^2$ (= first-order noise) region. Laeven et al. [18] derived the matching autocovariance function:

$$R_{xx}(\tau) = c\pi[E_1(\tau/t_n) - E_1(\tau/t_m)] \quad (24)$$

where E is an exponential integral, defined by:

$$E_1(\xi) = \int_{\xi}^{\infty} (e^{-s}/s)\, ds$$

4 Simple Data Processing

Neglecting peak height determination, the problems in simple chromatographic data processing, i.e. peak area and peak top determination, are how to find the peak start, the peak top and the peak end, how to correct for a drifting baseline, and how to determine the uncertainty in the results, particularly the peak area. Peak start, peak top and peak end are generally determined by calculating the first and second derivative of the peaks. Threshold levels, as close to zero as is permitted by the noise, are used to determine start, stop, top and valleys of single peaks or combinations of peaks. Several methods are used to correct for unresolved peaks and a drifting baseline.

Mostly, unresolved peaks are corrected by using a perpendicular dropline, a valley to valley correction or a combination. Simple correction methods, both for unresolved peaks and a drifting baseline, generally give systematic errors in the results. Another cause of errors is the limited number of data points of a digitized signal. Theoretically, about 7 or 8 equally spaced data points are sufficient to describe an approximately Gaussian peak. An appropriate interpolation procedure has to be used to determine the value between the points. However, a simple differentiation routine may cause big errors, particularly in case of unresolved peaks. Taking the nth derivative of a signal implies multiplying with $(j\omega)^n$ in the frequency domain:

$$\frac{d^n f}{dt^n} \leftrightarrow (j\omega)^n F(\omega) \tag{25}$$

This may cause severe problems with differentiating noisy signals, the high frequency components of the noise are amplified. In practice, differentiation has to be combined with a smoothing (low pass filtering) procedure.

An example is the relatively simple moving average filter. In case of a digitized signal, the values of a fixed (odd) number of data points (a window) are added and divided by the number of points. The result is a new value of the center point. Then the window shifts one point and the procedure, which can be considered as a convolution of the signal with a rectangular pulse function, repeats. Of course, other functions like a triangle, an exponential and a Gaussian, can be used.

Very popular is the Savitzky-Golay filter [20]. As the method is used in almost any chromatographic data processing software package, the basic principles will be outlined hereafter. A least squares fit with a polynomial of the required order is performed over a window length. This is achieved by using a fixed convolution function. The shape of this function depends on the order of the chosen polynomial and the window length. The coefficients b_i of the convolution function are calculated from:

$$\frac{\partial}{\partial b_k} \sum_{i=-m}^{+m} ((f_i - y_i)^2) = 0 \qquad y_i = \text{observed values} \tag{26}$$

where the window length is $2m + 1$.
The values:

$$f_i = \sum_{k=0}^{n} b_k i^k$$

are the calculated (smoothed) values.

Typical advantages of the Savitzky-Golay method are:

- the method can be used on-line
- it is possible to calculate a convolution function for determining the (smoothed) derivatives of the data set.

Data points at the beginning and the end of the data set are lost. Proctor and Sherwood [20] suggest an extrapolation with a polynomial of the same order, permitting repeated smoothing of one data set. Coefficients of the convolution function are given in [20] and, improved, in [21].

5 Uncertainty in Area Determination

The uncertainty in the determination of the area of a peak caused by (stationary) noise can be calculated both in the time domain and in the frequency domain. In [16] a derivation of σ_I^2, i.e. the variance of the integrated noise, is given:

$$\sigma_I^2 = 2 \int_0^T (T - |\tau|) \, R(\tau) \, d\tau \tag{27}$$

where
T = integration time
$R(\tau)$ = autocovariance function of the baseline noise.
The variance in the frequency domain reads:

$$k_{\overline{\sigma_I^2}} = 2T \int_0^\infty G(\omega) \frac{\sin^2(\omega T/2)}{(\omega T/2)^2} \, d(\omega T/2) \tag{28}$$

where $G(\omega)$ is the PSD of the baseline noise.

The bar and index are usual in annotating an ensemble average. As an example, σ_I^2 is calculated with Eq. (27) for first-order noise, i.e. white noise bandlimited by a first-order low pass filter:

$$\sigma_I^2 = \sigma_n^2 \left[2TT_1 + 2T_1^2 \left\{ \exp\left(-\frac{T}{T_1}\right) - 1 \right\} \right] \tag{29}$$

σ_n = standard deviation of the baseline noise.
In chromatography a part of this eqn can be neglected because $T_1 \gg T$ and reduces to:

$$\sigma_I^2 = \sigma_n^2 2TT_1 \tag{30}$$

The standard deviation is:

$$\sigma_I = \sigma_n \sqrt{2TT_1} \tag{31}$$

Equation (31) implies that the uncertainty of the determination of a peak area is proportional to the square root of the integration time (peak width). It should be noticed that in Eq. (31) σ_n and T_1 are not independent; σ_n is the standard deviation of the (first order) baseline noise for a given time constant T_1, determining the cut-off frequency of the noise. It can be proven [19] that the variance σ_I^2 is proportional to T^2 in case of 1/f noise.

The question arises which time constant is optimal with respect to the uncertainty. Of course, increasing the value of the time constant T_1 reduces the variance σ_n^2; on the other hand, the peaks will be distorted. The peak width is increasing and the peak height is decreasing with increasing T_1, as is known from the EMG model. Increasing peak width means increasing integration time, resulting in an increasing variance σ_I^2. It has been proven that the latter effect dominates. In practice the time constant has to be chosen as small as possible; low pass filtering is not advisable. It is important to notice that this is not true in case of peak start/peak stop determination procedures.

Laeven and Smit [22] presented a method for determining optimal peak integration intervals and optimal peak area determination on the basis of an extension of the mentioned theory. Rules of thumb were given, based on the rather complicated theory. Moreover, a simple peak-find procedure was developed, based on the derived rules.

6 Univariate Data Processing

The use of pre-information is in general typical for more advanced data processing. Of course, some pre-information is used in simple data processing, particularly for the determination of the peak integration start and stop. However, knowledge of the signal (peak shape) and noise characteristics is not used and thus the information extraction from the signal is not optimal. Even data processing with a computer is often far from optimal and the present data handling procedures are mostly based on relatively simple algorithms. Processing a "difficult" chromatogram involves solving the following problems: baseline drift correction, peak-find procedures (resolved and unresolved), peak parameter estimation in the presence of noise, and peak parameter estimation for unresolved peaks.

Hippe et al. [23] discussed numerical operations for computer processing of (gas) chromatographic data. Apart from a baseline correction method, a method of recognition of peaks is described. The relationship between the convexity of an isolated peak and the monotonic nature of its first derivative is used to find the most probable deflection points. The number of maxima and shoulders are used for a decision if the segment of the chromatogram contains an isolated peak or an unresolved peak complex. The number of shouders and maxima determine the total number of component peaks.

Scheeren et al. [7] describe peak shaped functions as an orthogonal polynomial series (Hermite), which means simultaneously low filtering, if the number of terms is limited. For example, calculating the first coefficient of the (modified) Hermite series implies determining the cross-correlation of the analytical signal with a Gauss function. This is equivalent to putting the analytical signal through a Gaussian filter, which can be considered as a matched filter for Gaussian peaks if this filter is adapted to the peak width of the peak. The scaling may be a problem; if it is not correct, than the results may be poor. A typical example of a Chebyshev polynomial approximation of a part of a chromatogram is shown in Fig. 6.

Curve fitting and peak deconvolution procedures with non-linear regression methods have been applied several times with varying success. The non-linear regres-

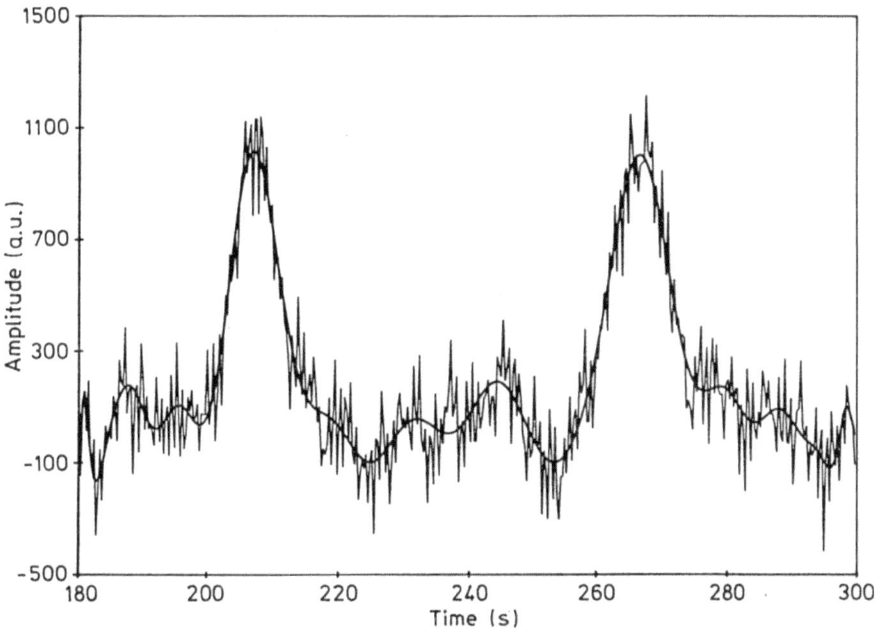

Fig. 6. Chebyshev approximation used as a filter on part of a chromatogram (0.1 µg l^{-1} anthracene, 0.5 µg l^{-1} benzanthracene, reversed-phase HPLC) 40 terms Chebyshev polynomial approximation.

sion is based on a least-squares fit of a mathematical model, non-linear in the parameters. The goodness of fit is defined by:

$$\chi^2 = \sum_{i=1}^{m} \left\{ \left(\frac{1}{\sigma_i^2} \right) [y_i - y(x_i)] \right\}^2 \tag{32}$$

y_i = measured data points
$y(x_i)$ = data points calculated from the mathematical model
σ_i^2 = variance of the data points y_i

If the function $y(x_i)$ with the parameters a_j ($j = 1, n$) is considered, it is possible to minimize χ^2 for all a_j simultaneously. The n-dimensional parameter space has to be searched for the minimum χ^2 value. In the gradient-search of least squares, all the parameters a_j are incremented simultaneously. The relative magnitudes are adjusted in such a way that the resulting direction of travel in parameter space is along the direction of maximum variation of χ^2. The popular Marquardt algorithm combines the best features of gradient search with a method of linearizing the fitting function. One of the problems with non-linear regression techniques is to find the absolute minimum in the hyper-surface and to recognize and avoid so-called local minima. A good estimation of the initial values, particularly the number of peaks to be expected, is essential. The hyper-surface must be limited by estimating confidence intervals and absolute limits of the initial parameters; using prior knowledge increases the chance to find the absolute minimum.

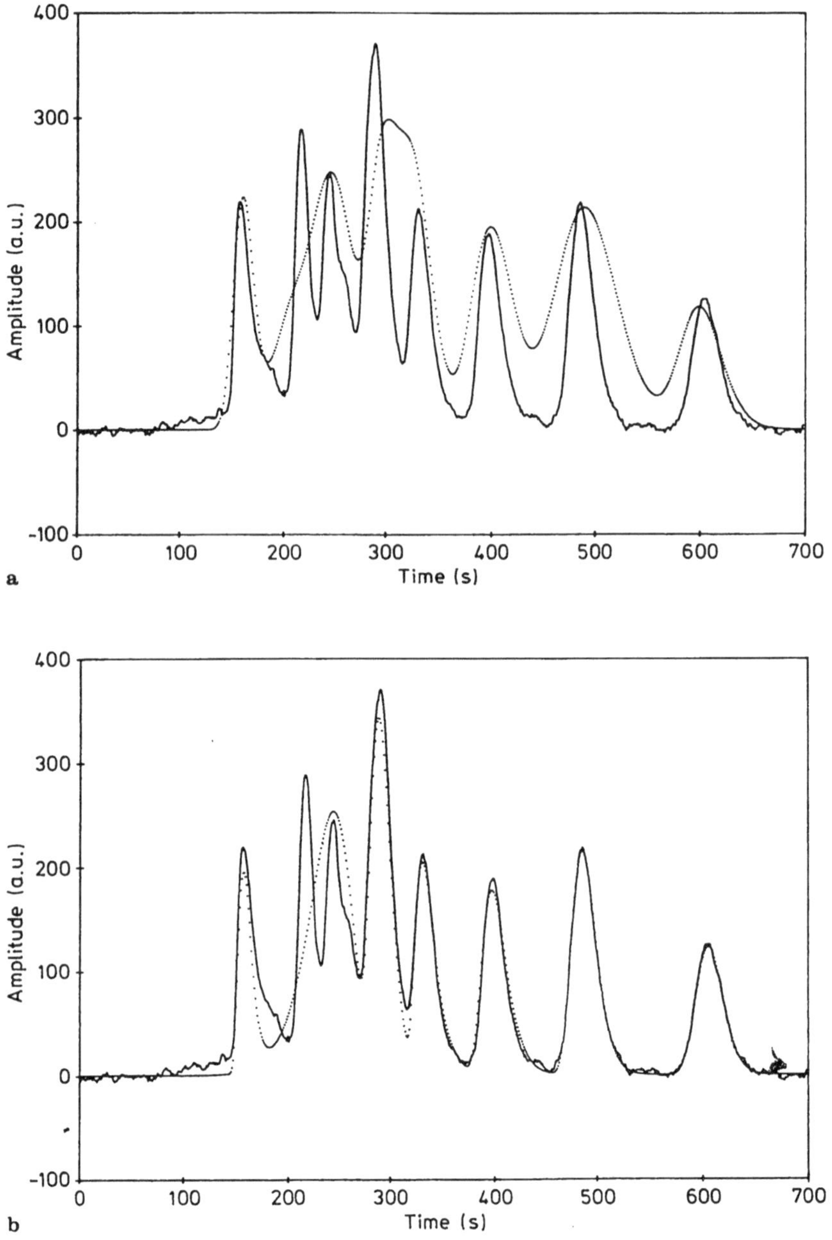

Fig. 7. Chromatogram of alkylbenzenes (solid lines, see Fig. 6). Dotted lines: **a)** data points calculated from the initial values of the parameters using a regression model with 8 Fraser-Suzuki functions without baseline; **b)** data points calculated after three regression cycles; **c)** data points calculated after 47 regression cycles; **d)** regression model extended with 3rd order baseline after 47 cycles.

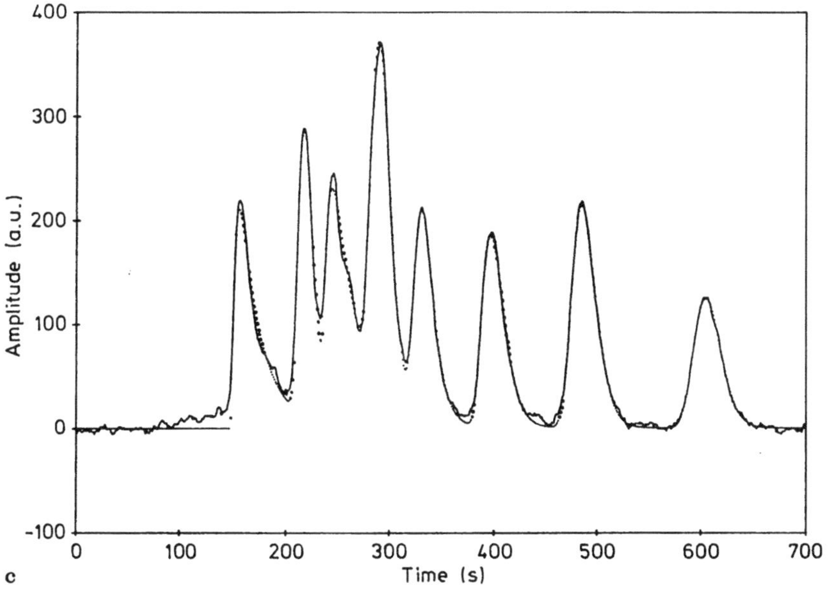

c

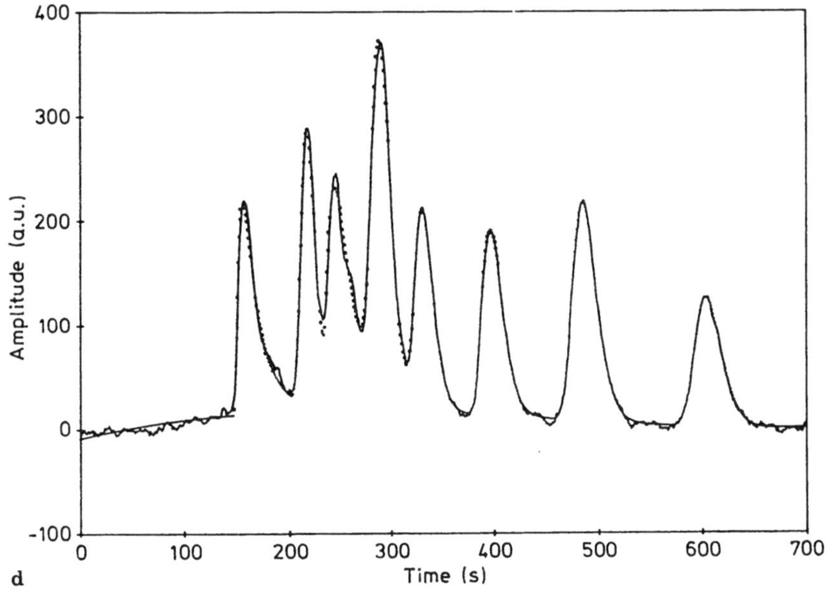

d

A general usable non-linear regression software package for evaluation of peak parameters is described in [15]. A complete chromatogram, including a drifting baseline, can be fitted with suitable models for the peaks and the baseline (Fig. 7). This estimation method, using the whole chromatogram, is probably the optimum way of correcting the influence of a drifting baseline.

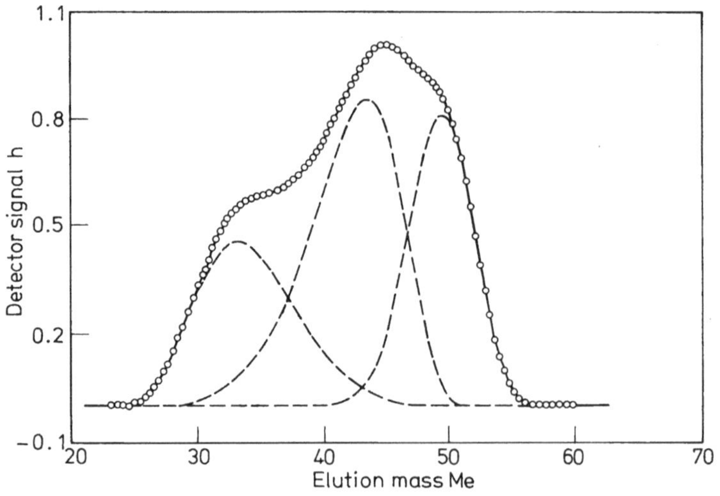

Fig. 8. Deconvolution of a chromatogram of Dextran T-2000 with water as the mobile phase in a glyceryl-coated controlled porous glass column.

Vardya and Hester [14] used their OEX model in a constrained linear optimization procedure, based on the Box-Complex method which is essentially a constrained simplex minimization technique. The method does not require derivatives of the object function and is not subject to scaling problems. As an example, Fig. 8 shows a deconvolution of a chromatogram of Dextran T-2000 with water as the mobile phase in a controlled porous glass column. The badly fused peaks are successfully deconvoluted.

7 Multichannel Data Processing

In general, chromatographic methods offer high resolution and selectivity. Nevertheless, partially separated compounds in a mixture regularly occur, resulting in incomplete resolution of chromatographic peaks. Single channel detectors, for example a flame ionisation detector (FID) in GLC or a single wavelength UV detector in HPLC, do not contribute much to extracting relevant information from a cluster of unresolved peaks; chemometric methods like the previously described non-linear regression method, have to be applied to estimate the relevant analytical signal parameters.

However, combining chromatographic separation with multichannel detection offers a much more powerful technique for the quantitative and qualitative analysis of mixtures; an extra dimension is added to the measurement. Examples of chromatographic techniques which employ multichannel detection are GC/MS (gaschromatography/mass spectrometry), HPLC (high performance liquid chromatography)/ multi wavelength detection (diode array), and GC/IR (infrared spectrometry). The extra dimension contributed by the multichannel detector can be utilized in different ways and with different objects in view.

From the point of view of data processing, however, the utilized methods of data processing are related and often based on the same mathematical principles. For example, about the same procedures can be used for both GC/MS and HPLC/multi-wavelength detector combinations. In the simplest case the constituents can be completely "resolved" by the detector by, for instance, selecting a specific wavelength for each component. In general, methods like pattern recognition and related techniques, e.g. feature selection, spectrum matching, and library search, are applied if the extra dimension is very informative (MS, FTIR) and they can serve for identification of the eluting constituents. But often the additional information has to be supplied by some kind of chemometric process like multivariate curve resolution, because the components of the mixture are overlapping both chromatographically and spectrally.

Deconvolution methods may be arranged in different ways; for instance by final object, the nature of the pre-information, or the applied basic mathematical theory.

Objects are: A_1. Peak purity test.
A_2. Determination of the number of components.
A_3. Resolution in the time domain.
A_4. Determination of spectra of the components.

Pre-information: B_1. Number of components.
B_2. Spectral pre-knowledge.
B_3. Elution pre-knowledge.
B_4. Remaining pre-knowledge.

The mathematical techniques are part of multivariate statistics. They are closely related and often exchangeable. Two main approaches can be distinguished: Least Squares Optimization (LSO), and Factor Analysis (FA).

8 Basic Multivariate Analysis

A data matrix [D] can be considered as the product of two matrices [A] and [C]:

$$[D] = [A]\,[C]$$

[A] could represent the spectral information and [C] the chromatographic (time dependent) information. Assumed is that the mutual influence of the components is neglectable, i.e. that the system including the detector is linear. In principle only [A] has to be determined if a model exists for the columns of [C]:

$$[A] = [D]\,[C]^{-1} \tag{33}$$

However, one has to keep in mind that [C] is not really known, only a model exists. Now [A] can be estimated with the generalized inverse of the model matrix:

$$[C]_{\text{gen. inv.}} = \{[C]_{\text{mod}}^{T}\,[C]_{\text{mod}}\}^{-1}\,[C]_{\text{mod}}^{T} \tag{34}$$

$$[A]_{\text{est}} = [D]\,[C]_{\text{gen. inv.}} \tag{35}$$

$[A]_{est}$ is the least squares estimation of [A], given the model of [C], and an estimation of [D] can be calculated:

$$[D]_{est} = [A]_{est} [C]_{mod} \tag{36}$$

The model $[C]_{mod}$ can be evaluated by comparing $[D]_{est}$ with [D] using a χ^2-criterion. The parameters of the model can be optimized by some optimization procedure (Simplex).

In the Factor Analysis method [D] is subdivided into two matrices, applying Principal Component Analysis (PCA). This is a pure mathematical solution. There is no relation between the obtained matrices and the physical parameters. However, it is possible to relate the abstract solution resulting from the PCA to a physical relevant solution, using some pre-knowledge.

Target Testing, for instance, is a procedure with the goal to rotate the abstract solution to a model factor. If this is not possible, the model is not correct.

9 Applications of Multivariate Techniques

One of the first applications of LSO in multichannel processing (GC/MS) is published by Knorr et al. [24]. An EMG model is assumed for the elution profiles, where all model parameters except the retention time t_R are determined in advance (B_3). They achieve the objects A_2, A_3 and A_4 in case of clusters of two and three components.

King and King [25] extend the method of [24] by using a more complicated elution model. All parameters are determined with LSO, which implies that no standards have to be determined. Assumed is a pre-knowledge of the number of components (B_1), an exponential down scan correction (B_4), a background subtraction (B_3), and a saturation correction (B_2).

It should be noted that the generalized inverse of $[C]_{mod}$ is determined with PCA. The inversion step, which is risky in case of badly resolved peaks because of the probability of an ill-conditioned matrix [C], is circumvented.

Another extension is given by Frans et al. [26]. This method is adapted to LC-UV. The number of components (5 to 8) is determined with PCA. Lindberg et al. [27] apply a LSO still more close to FA than the method applied by King and King, and they claim a better accuracy for their Partial Least Squares (PLS) method compared to real LSO or FA.

Sharaf and Kowalski [28] are using the pre-knowledge of non-negativity of the spectra (B_2) and positive additivity of each linear combination yielding measurement points for the rotation of the abstract PCA solution to a physically relevant solution. The result is a solution-band for their GC/MS problem. The bandwidth depends on the resolution, the concentration ratio of the components and the correlation in the component spectra. The method is adapted to LC-UV by Osten and Kowalski [29]. Only binary systems are investigated, resulting in both A_3 and A_4. Vandeginste et al. [30] propose a method analogous to the Sharaf/Kowalski procedure. Their method, applied to LC-UV, permits two approaches: using spectral pre-knowledge (non-negativity, positive linear combinations, single spectra with the smallest area to norm ratio as the best estimation of the pure spectra) or using chromatographic pre-knowledge (non-negativity, single-maximum, minimal area). The result is not a solution band but a unique solution. Two and three component systems are successfully deconvoluted. Both elution profiles and spectra can be determined. The number of components is not determined.

Most pre-knowledge is used by McCue and Malinowski [31]. Target testing is applied to look for certain components in the cluster, where the spectrum of the pure component is the target to be tested. If the targets of the cluster together are explaining the data, then the number of components is known. The method can be considered as a kind of library search action.

Iterative Target Testing is another approach. The preliminary approximations of the real factors are chosen, based on the first (VARIMAX) rotation of the abstract PCA solution. With iterative target testing the factors are transformed to the best approximations. It can be considered as LSO, where PCA and VARIMAX are supplying the model. Clusters with an arbitrary number of peaks can be deconvoluted. Six component systems are tested (Vandeginste et al. [32]).

10 Information from Chromatographic Data Sets

In a chromatographic separation procedure the parameters of the chromatographic system (stationary phase, flow, temperature, etc.) have to be selected respectively optimized with respect to some criterion (resolution, time, etc.). In gas chromatography retention data series are published [33, 34] and used for the study of solvent/solute interaction, prediction of the retention behaviour, activity coefficients, and other relevant information usable for optimization and classification. Several chemometric techniques of data analysis have been employed, e.g. PCA, numerical taxonomic methods, information theory, and pattern recognition.

Pattern recognition can be applied for the determination of structural features of unknown (monofunctional) compounds (Huber and Reich [35]). The information about the chemical structure is contained in a multidimensional gas-liquid retention data/stationary liquid phases set. The linear learning machine method is applied in a two step classification procedure. After the determination of a correction term, the skeleton number, a classification step for the determination of the functional group is executed. It is remarkable that 10 stationary phases are sufficient for the classification.

Pattern recognition methods can also be employed for the classification of stationary phases and quantification of their retention characteristics (Huber and Reich [36]). The large number of stationary phases can be drastically reduced to a standard set of solvents which have significantly dissimilar retention characteristics. Several selection criteria are applied:

1) The polarity number, suggested by McReynolds [37].
2) The euclidian distance (ED) of the selected phases relative to the most non-polar stationary phase, i.e. squalane.

   The ED defined for two solvents, p and r, is:

$$d_{pr} = \sqrt{\sum_{i=1}^{k} (x_{ip} - x_{ir})^2}$$

3) The mean of the retention indices (MI) of all key solutes.
4) The length obtained by the traversal of the minimum spanning tree (MST).

The results of the feature reduction process and classification are extensively tabulated in [36].

Predicting chromatographic retention data can be successfully done by numerical taxonomic aggregation and FA [38,39,40]. Both techniques were employed before in chromatographic data analysis, for example the prediction of activity coefficients and for the identification of principal components with strong influence on the retention behaviour, respectively.

Cluster analysis (numerical toxonomic aggregation) [38,40] is applied to arrange phases according to their chromatographic behaviour. A set of retention data for 16 monofunctional benzenes, 110 difunctional benzenes and 15 trifunctional benzenes was subjected to analysis. Three groups of stationary phases can be distinguished: polar, non-polar, and polyfluorinated. A linear relationship between the retention data of two stationary phases of the same class can be worked out. This linear relationship fits the model

$$\log t'(\varphi Z, \Phi_1, T_1) = \alpha \log t'(\varphi Z, \Phi_2, T_2) + \beta$$

where $\log t'$ is the net retention time, φZ is the solute, Φ_1 and Φ_2 are the stationary phases, and T_1, T_2 are the temperatures.

Factor analysis (PCA) is applied to a retention time data set of 17 benzenic monosubstituted compounds on 21 stationary phases [39,40]. The result is the relationship:

$$x_{P,\varphi_i} = a_\varphi \cdot x_{P,\varphi_1} + b_\varphi \cdot x_{P,\varphi_2} + c_\varphi \cdot x_{P,\varphi_3} + d_\varphi$$

where x_{P,φ_n} ($n = 1, 2, 3$) are the experimental data of a substance P, measured on three phases φ_1 (non-polar), φ_2 (polyfluorinated), and φ_3 (polar).
a_φ, b_φ, c_φ, and d_φ are the coefficients of a non-restricted multiparametric relationship.

11 Special Techniques

The previously described analysis and process techniques are applied to chromatographic data and signals obtained from conventional chromatographic systems. However, the introduction of chemometric techniques, particularly correlation procedures, permits a different approach to chromatographic data analysis and data processing. Correlation chromatography (CC) is an example of a modified chromatographic technique, based on a system-theoretical approach and a rather complicated computerized data processing.

In conventional chromatography the chromatogram is the response of the chromatographic system on an impulse-shaped single injection or sample. Correlation chromatography, however, utilizes semi-continuous multiple random injections of sample over a period of time. The resulting random response of the system is cross-correlated with the used input function. The correlogram is identical to the chromatogram obtained from a single injection. If the chromatographic system is contaminated with

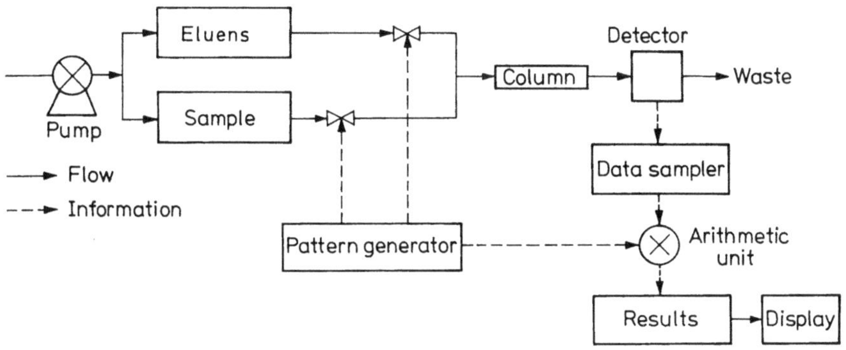

Fig. 9. Flow sheet for a correlation chromatography system.

noise, this noise is not correlated with the input, and its contribution to the overall cross-correlation function converges to zero with increasing correlation time. A considerable improvement of the signal to noise ratio can be achieved in a relatively short time. Figure 9 shows a set up of a correlation chromatograph [41].

Comparing correlation chromatography and conventional chromatography shows that there are no differences concerning the columns and detectors. The injection system has to be modified and a (micro)computer based "correlator" for data processing and generation of a suitable random pattern has to be added. Mostly in CC the sample is semi-continuously introduced according to a (pseudo) random pattern, a so-called pseudo random binary sequence (PRBS). In a random binary pattern only

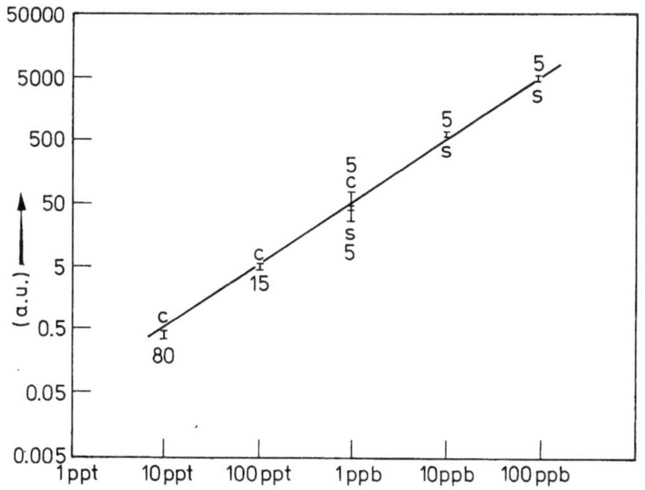

Fig. 10. Calibration graph for phenol with fluorimetric detection for 5 concentrations (10^{-4} to 10^{-8} g l^{-1}); s indicates single injection and c indicates correlation chromatography. The numbers below the data points indicate the correlation time (minutes) (ppt = ng l^{-1}). The bars represent $\pm 3 \times$ standard deviation of the integrated noise (confidence interval).

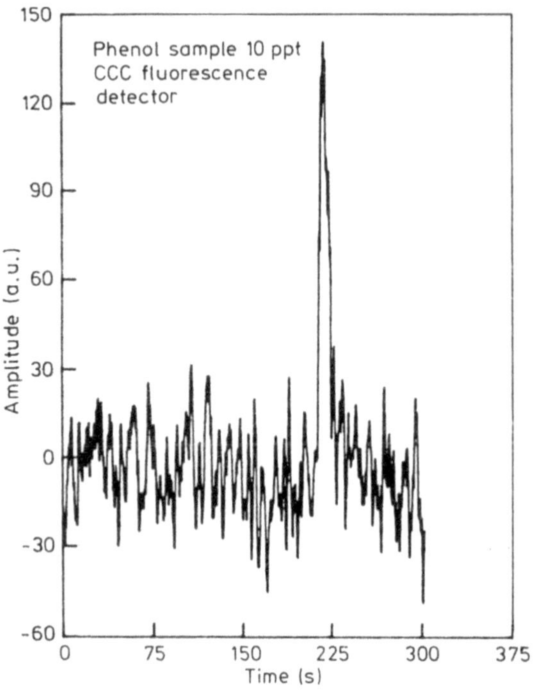

Fig. 11. Correlogram of a 10^{-8} g l^{-1} phenol sample. Detection limit is approximately 3 ppt (3.10^{-9} g l^{-1}).

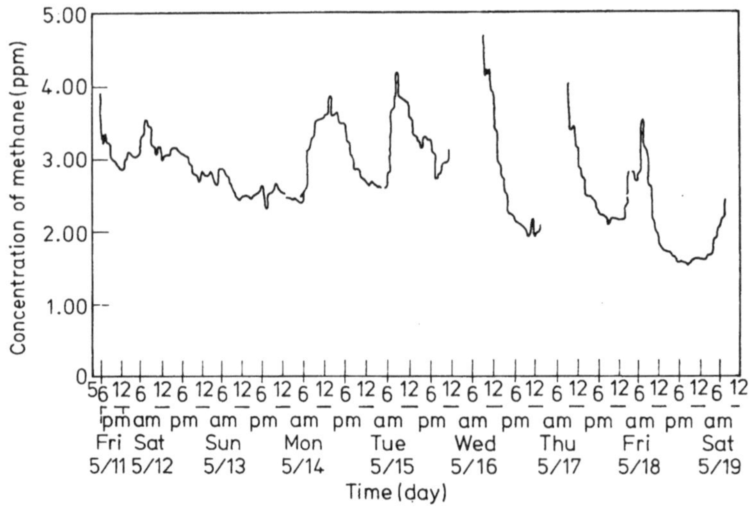

Fig. 12. Profile of the concentration of methane in the ambient air showing 1-h averages for an 8-day period. Determined with correlation (multiplex) chromatography utilizing catalytical oxidation as a modulation.

two signal values, e.g. 0 and 1, are possible; which of the two will be present at any particular time is not predictable. However, a PRBS is a logical function combining the properties of a true (binary) random signal with those of a reproducible deterministic signal. After a certain time, a sequence, the pattern is repeated. The value 0 of the PRBS corresponds with the flow of pure mobile phase into the column; if the value is 1, sample is injected.

Researchers are active in the field of correlation gaschromatography and correlation HPLC [41–47], the first application in trace analysis was introduced in 1970 [48]. A typical example of the noise reduction property is the determination of a calibration graph of phenol for the higher concentrations with conventional chromatography, and extended to very low concentrations by CC (Fig. 10). The detection limit achieved is about 3 ppt (Laeven et al. [46]). A correlogram of 10 ng/l phenol sample is shown in

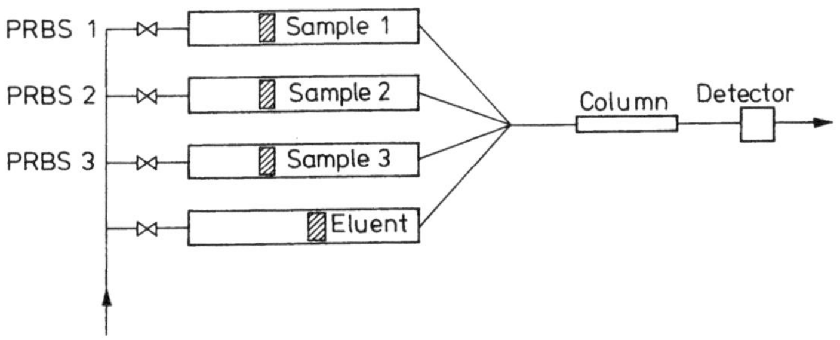

Fig. 13. Set-up of a simultaneous HPLC system.

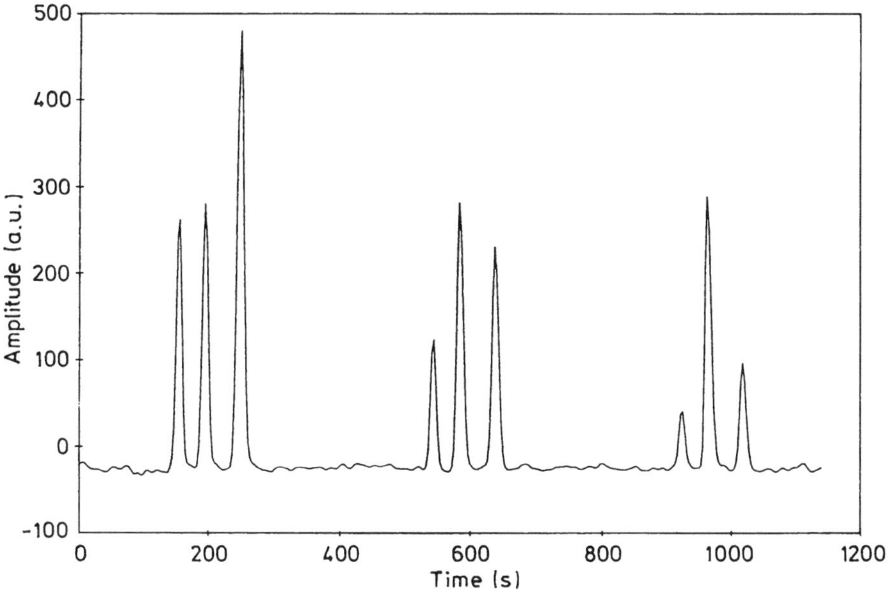

Fig. 14. Simultaneous chromatogram of three samples, each containing naphthalene, anthracene and 1,2 benzanthracene with different concentrations.

Fig. 11. The mechanical valves can be replaced by thermal or chemical modulating devices. In principle, this modification circumvents the need of reliable mechanical valves and offers a more flexible system [49,50]. An example is the thermal oxidative modulation of methane in air. Figure 12 shows a concentration profile during 8 days, determined with an experimental modulation CC set up [50].

An extension of CC is Simultaneous Correlation Chromatography (SCC) (Smit et al. [51]); Fig. 13 shows an experimental set up. Three samples with naphthalene, anthracene and 1,2-benzanthracene are simultaneously injected, however, each controlled by a sequence uncorrelated with the others. The result is shown in Fig. 14. The peaks of naphthalene are used to construct a calibration line. The advantages are twofold: The random fluctuations are reduced by multiple injection and averaging property, and both an unknown sample and calibration samples are measured simultaneously under exactly the same conditions, drift and uncertainty are reduced to a high extent.

Finally, CC can be used for monitoring of processes and reactions. The semi-continuous nature of CC permits an almost continuous monitoring of a process in contrast with conventional chromatography. The technique is applied in pyrolysis GC (Py-GC), a valuable method for characterizing materials. The determination of the kinetics is difficult because of the bad reproducibility of the degradation temperature and sample volume. The application of a linear temperature increase with continuously analyzing samples is preferable, but hardly practicable. CC is a good alternative, as is proved by Kaljurand and Küllik [52]. Figure 15 shows correlograms obtained from the degradation products of polycaproamide.

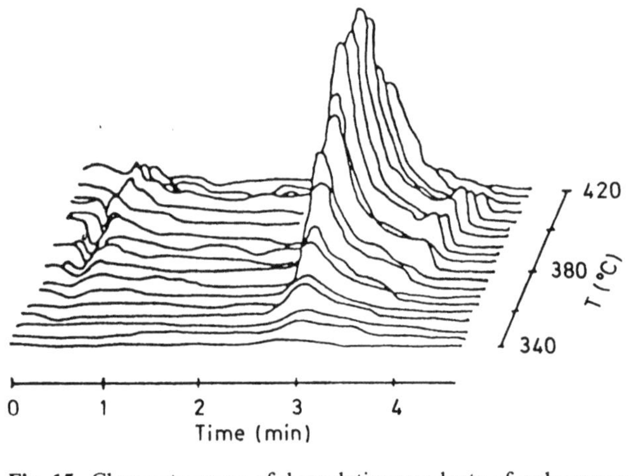

Fig. 15. Chromatograms of degradation products of polycaproamide by pyrolysis correlation chromatography.

12 References

1. Smit, J. C.: Thesis, University of Amsterdam 1981
2. Smit, J. C., Smit, H. C., de Jager, E. M.: Anal. Chim. Acta *122*, 1 (1980)
3. Smit, J. C., Smit, H. C., de Jager, E. M.: ibid *122*, 151 (1980)

4. Rhee, H. K., Aris, R., Amundsen, N. R.: Phil. Trans. of the Royal Soc. *A 267* (1970)
5. Lankelma, J. V., Smit, H. C.: Communications in Mathematical Chemistry *17*, 157 (1985)
6. Grubner, O.: Anal. Chem. *43*, 1934 (1971)
7. Scheeren, P. J. H., Klous, Z., Smit, H. C.: Anal. Chim. Acta *171*, 45 (1985)
8. Debets, H. J. G., Wijnsma, A. W., Doornbos, D. A., Smit, H. C.: ibid. *171*, 33 (1985)
9. Grushka, E.: Anal. Chem. *44*, 1733 (1972)
10. Littlewood, A. B., Anderson, A. H., Gibb, T. C.: J. Chrom. Sci. *8*, 640 (1970)
11. Yau, W. W.: Anal. Chem. *49*, 395 (1977)
12. Fraser, R. D. B., Suzuki, E.: ibid. *41*, 37 (1969)
13. Smit, H. C., Smit, J. C., de Jager, E. M.: Chromatographia *22*, 123 (1986)
14. Vaidya, R. A., Hester, R. D.: J. Chromatogr. *287*, 231 (1984)
15. Scheeren, P. J. H., Barna, P., Smit, H. C.: Anal. Chim. Acta *167*, 65 (1985)
16. Smit, H. C., Walg, H. L.: Chromatographia *8*, 311 (1975)
17. Duursma, R. P. J., Smit, H. C.: Anal. Chim. Acta *133*, 67 (1981)
18. Laeven, J. M., Smit, H. C., Lankelma, J. C.: ibid. *157*, 273 (1984)
19. Smit, H. C., Walg, H. L.: Chromatographia *9*, 483 (1976)
20. Savitzky, A., Golay, M. J. E.: Anal. Chem. *36*, 1627 (1964)
21. Proctor, A., Sherwood, P. M. A.: ibid. *52*, 2315 (1980)
22. Laeven, J. M., Smit, H. C.: Anal. Chim. Acta *176*, 77 (1985)
23. Hippe, Z., Bierowska, A., Pietzyga, T.: ibid. *122*, 279 (1980)
24. Knorr, F. J., Thorsheim, H. R., Harris, J. M.: Anal. Chem. *53*, 821 (1981)
25. King, M. D., King, G. S.: ibid. *57*, 1049 (1985)
26. Frans, S. D., McConnell, M. L., Harris, J. M.: ibid. *57*, 1552 (1985)
27. Lindberg, W.. Öhman, J., Wold, S.: ibid. *58*, 299 (1986)
28. Sharaf, M. A., Kowalski, B. R.: ibid. *53*, 519 (1981)
29. Osten, D. W., Kowalski, B. R.: ibid. *56*, 991 (1984)
30. Vandeginste, B., Essers, R., Bosman, T., Reyen, J., Kateman, G.: ibid. *57*, 97 (1985)
31. McCue, M., Malinowsky, E. R.: Appl. Spectrosc. *37*, 463 (1983)
32. Vandeginste, B., Derks, W., Kateman, G.: Anal. Chim. Acta *173*, 253 (1985)
33. Rohrschneider, L.: J. Chromatogr. *22*, 6 (1966)
34. McReynolds, W. O.: Gas Chromatographic Retention Data, Preston Technical Abstracts Company, Evanston, Illinois, USA 1966
35. Huber, J. F. K., Reich, G.: Anal. Chim. Acta *122*, 139 (1980)
36. Huber, J. F. K., Reich, G.: J. Chromatogr. *294*, 15 (1984)
37. McReynolds, W. O.: J. Chrom. Sci. *8*, 685 (1970)
38. Fellons, R., Lafaye de Micheaux, D., Lizzani-Cuvelier, L., Luft, R.: J. Chromatogr. *213*, 223 (1981)
39. #: Ibid, *248*, 35 (1982)
40. Fellons, R., Lizzani-Cuvelier, L., Luft, R.: Anal. Chim. Acta *154*, 191 (1983)
41. Smit, H. C., Duursma, R. P. J., Steigstra, H.: ibid. *133*, 283 (1981)
42. Annino, R.: J. Chromatogr. *14*, 265 (1976)
43. Phillips, J. B.: Anal. Chem. *52*, 468A (1980)
44. Annino, R., Gonnord, M. F., Guiochon, G.: ibid. *51*, 379 (1979)
45. Smit, H. C., Lub, T. T., Vloon, W. J.: Anal. Chim. Acta *122*, 267 (1980)
46. Laeven, J. M., Smit, H. C., Kraak, J. C.: ibid. *150*, 253 (1983)
47. Kaljurand, M., Kullik, E.: Chromatographia *11*, 328 (1978)
48. Smit, H. C.: ibid. *3*, 515 (1970)
49. Phillips, J. B., Lun, D., Pawliszyn, J. B.: Anal. Chem. *57*, 2779 (1985)
50. Valentin, J. R., Carle, G. G., Phillips, J. B.: ibid. *57*, 1035 (1985)
51. Smit, H. C., Mars, C., Kraak, J. C.: Anal. Chim. Acta *181*, 37 (1986)
52. Urbas, E., Kaljurand, M., Kullik, E.: J. Anal. and Appl. Pyr. *1*, 213 (1980)

Chemometrics in Food Chemistry

Michele Forina, Silvia Lanteri and Carla Armanino

Istituto di Analisi e Tecnologie Farmaceutiche ed Alimentari
Via Brigata Salerno, ponte 16147 Genova, Italy

Table of Contents

1 Introduction

During the last twenty years, food chemists have been using an increasing number of analytical instruments to analyse several samples quickly and obtain, in a short time, a great deal of chemical information from each sample. At the same time, they have increased their knowledge of the chemical composition of natural foods and of the changes due to storage and treatments, and also of market and customer requirements.

Long before, experience had already shown that such close relationships exist between some chemical (or physical) variables describing food composition that, for some characterizations, a combination of chemical or physical quantities is more meaningful than each quantity alone. Consequently, sums, ratios or ratios of sums of chemical quantities have been (and are being) studied to characterize food quality, origin or treatments. This search for useful features has been limited to combinations of a small number of variables, generally two, because of the limitations of the available computational facilities, until a few years ago.

Nevertheless, these studies show a development towards a multivariate point of view, as the high number of chemical parameters produced by modern analytical instrument requires. The spread of computing systems of all sizes and the circulation of their respective software have allowed the great amount of computing deriving from the large number of measured variables and their relationships to be dealt with effectively.

The arrival of computers in every chemical laboratory has made possible the use of multivariate statistical analysis and mathematics in the analysis of measured chemical data. Sometimes, the methods were inadequate or only partially suitable for a particular chemical problem, so handling methods were modified or new ones developed to fit the chemical problem. On the basis of these elements, common to every field of chemistry, in 1974 a new chemical science was identified: chemometrics, the science of chemical information. In the same year, Bruce Kowalski and Svante Wold founded the Chemometrics Society, which since then has been spreading information on multivariates in chemistry all over the world.

Chemometrics is the chemical science that uses mathematics, statistics and informatics:

a) to select or design optimal procedures and experiments;

b) to obtain the maximum useful information from the experimental chemical data. In food chemistry, chemometrical technique results are most necessary and promising, firstly because the chemical systems investigated in food chemistry are very complex and formed by many chemical species that are often very important even at trace level. The perception and experience of a researcher are not sufficient to single out the really significant information if these qualities are not suitably developed and supported.

Secondly, the study of the chemical composition of foods and of its changes on processing are not the only aim (except in control analyses): the space of chemical quantities is intermediate between the cause space, which includes every parameter affecting food composition, and the effect space, where we find variables related to food properties, its quality, sensorial evaluations, nutritional value, and storage possibilities. Generally, the chemical space is described accurately and concisely.

The best use of its information is essential to try to understand the relationships between cause and effect spaces.

This close connection and reciprocal necessity between food chemistry and chemometrics are shown by the several chemometrical methods being planned or further developed for these purposes or being tested in food chemistry problems.

Among these methods, multicalibration (multivariate calibration) [1,2] is important. Multicalibration is the final development of indirect analytical methods. The analytical method has been previously defined as the whole of operational steps (reactions, separations, . . .) that lead to a highly selective endpoint where one measured physical variable is univocally related to one chemical variable (quantity, concentration, . . .); this correlation is shown by the calibration curve (a straight line, generally). Multicalibration brings a complete change of this definition: the analytical method is the whole of chemical and mathematical operations that enable us to reach a multivariate selective system where several measured physical quantities are univocally related to several chemical quantities; the correlation is shown by the calibration hypersurface. Multicalibration is surely destined to be used with great effect in many areas in the future.

But, beside these elements of general interest, in food chemistry there is a lack of elementary knowledge of statistics, which points out how much ground is still to be covered in the diffusion of chemometrical methods and their correct application. Often, indeed, there is no diagram or description of the experiment, and sometimes samples are scarcely representative, as in some studies the easiest rather than a controlled or random sampling is used [3]. In many cases, the collection of samples suitably representing the system may be very hard or expensive, but this cannot justify the acrobatics. The limiting factors of the experimental design have to be shown in order to give information and suggestions for further research. By a suitable experimental design and chemometrical evaluation, the variance of analytical data due to the studied phenomena (not due to variations of method or laboratory) furnish very useful results.

2 Problems

Most of the problems solved by chemometrical methods concern

a) Description
b) Classification and modelling
c) Correlation
d) Clustering
e) Feature selection
f) Optimization

Experimental results are generally grouped in tables: two-dimensional matrices $\mathbf{X}_{NV}$ formed by N rows (objects = samples) and V columns (variables = chemical quantities, sensorial scores, physical quantities, . . .). It is very difficult to read and understand the information contained in a large data matrix, therefore it is really useless.

Graphic methods of representation are strongly recommended for displaying experimental results and for supervising the elaboration of data (Sec. 3.1). The usefulness of histograms and variable-by-variable plots is improved by using colour three-dimensional plots[4], by two-dimensional histograms, eigenvector projection (Sect. 3.2) and NLM plots.

2.1 Classification

In classification problems the data table is divided horizontally into two or more categories into which objects are grouped. The problem is to make the best use of the variables: to classify the objects into categories and, chiefly, to predict the category of objects of unknown category and to evaluate the correctness of this assignment.

The meaning of category is various in food chemistry: it may be a vegetable (or animal) species (e.g., corn, rice, oats), a cultivar of grape (e.g., Pinot wine, Tocai wine), the geographical origin or the process method. Only in a few examples the subdivision into categories is sharp, generally it is an oversimplification. Categories formed by different geographical origins of foods also include differences due to variety, soil, composition, climate and harvesting, storage and processing methods. For instance, the composition of the acid fraction of edible vegetable oils depends on climate. When two groups of samples are collected in two small areas of different climate, but homogeneous within each area, the groups define two categories.

The subdivision into categories of a data matrix corresponds to the introduction of another variable, discrete and integer: the category index. If the variable is really discrete, e.g., corn = 1, oats = 2, rice = 3, the subdivision into categories is true. If the variable is continuous, the subdivision into categories is true if it has the intracategory variance much smaller than the intercategory one. If the two variances are comparable, the subdivision into categories is not meaningful. Therefore, it is possible that a sample is considered "typical of a region" and a sample of a nearby area is by force considered "typical of another region" while it has the same characteristics as the first one.

Indeed, these categorizations are produced by the demands of the market, and by legislative and national needs. The chemometrician must check whether the measured data justify and make possible the categorization, because there are usually variables that are not always identified or measurable that cause changes of sample composition. The distribution of the samples in the space of these variables is not homogeneous but is in narrow ranges that define the related categories.

When the evident or ignored variables defining the categorization are continuous, the problem is instead a correlation problem. The classification methods are qualitative, while the correlation ones are quantitative.

2.2 Modelling

Modelling may be considered a particular case of classification, but really it has evolved from classification methods. In classification, the boundaries between every pair of categories are emphasized. In modelling, each category is considered alone and its model, its chemical-statistical description, is evaluated. The model boundaries may be considered as the line separating the category being studied from the wide category formed by all the samples of all the other categories. Modelling is the multivariate evolution of the tolerance intervals.

In Fig. 1 the ellipses show the category models at a determined confidence level. The ellipse projections on the axes, univariate confidence intervals, correspond to acceptance intervals. A sample with lowest or highest values of both variables follows univariate acceptances, but it can fall outside the model of the category, because the model is characterized by a correlation between the two variables, so that, for example, high values of both variables are (in the case of negative correlation) connected with a probability too low to allow a significant acceptance. So, the acceptance of a food product on the basis of the univariate acceptance intervals (being used to define its origin or toxicity, for example) may cause serious errors.

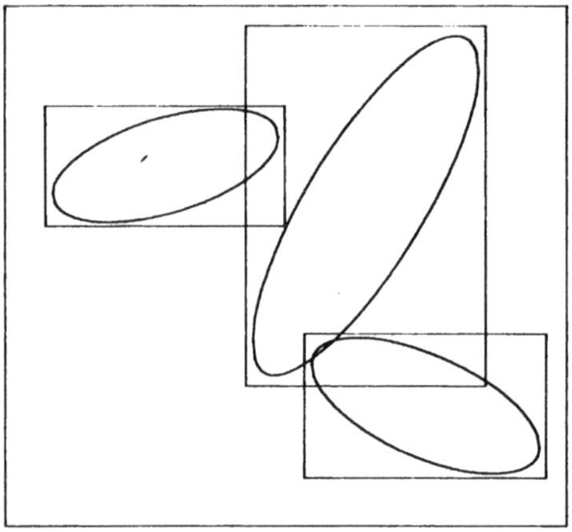

Fig. 1. Univariate vs. multivariate (bivariate) approach

After modelling, we return to classification with some more elements. Samples are univocally classified into a category if they are within the model of only one category; they are considered as belonging to two or more categories when the category models overlap and the samples are in the overlapping space (in this case it is possible to determine the probability of belonging to each category); the samples are outliers when they are within no category model. They are anomalous samples, they cannot be classified because they really belong to a category that is not present in the problem (the pure classification methods would instead place them in the least-different category).

2.3 Quantitative Chemometrics: Correlation

Correlation problems concern the study of the data tables of uncategorized objects, divided vertically; the vertical divisions correspond to two (or more) blocks of variables (block X, block Y, . . .). One or more variables may be in a block.

The greatest problems of these techniques concern the blocks of variables where block X is a prevalently chemical block and block Y consists of variables of different origin, sometimes mixed (e.g., geochemical and climatic).

X	Y
Chemical or mixed	Chemical or mixed
(chemical and physical)	Physical (instrumental)
	Geochemical
	Variety
	Climatic
	Alimentary (diet of animal)
	Cooking, storage
	Production, technology
	Sensorial
	Commercial
	Nourishment level

Obviously, some problems may have a block X that is nonchemical, e.g., in correlations between sensorial and commercial variables, but such studies are not included in this review. Generally, the correlations between variables of two blocks need different knowledge of different areas of study, only some of which may be considered chemical.

According to the type of block Y, the correlation can be used for different problems. The correlation between different chemical variables is used to investigate whether some quantities can be evaluated from others, so that the measurement of the first quantities can be avoided (major and minor chemical components, traces), or whether the relationships between the groups of chemical variables can be reduced (e.g., acid and sterolic fraction).

Multivariate calibration is based on the correlation between chemical and physical variables, it is the transformation of a lack of univariate selectivity into a multivariate one. The relationships between chemical variables and those producing food composition (geochemical, variety, alimentary, . . .) explain the changes caused by factors that cannot be changed one at a time or that have strongly interacting effects.

Knowing the relationships between chemical and sensorial variables, objective methods can be obtained to evaluate the food quality. Juries of experts cannot be formed and used so easily as the measurement of chemical quantities can. Besides, the knowledge of these relationships will be able to retain, so to speak, sensorial evaluations and follow the evolution of taste over a long period, so that it may be foreseen as well.

In an undifferentiated table, groups of similar objects or variables can be singled out by clustering techniques. These techniques can also be used to ascertain whether the subdivision into categories is exact, when they are applied to samples. When they are applied to variables, it is possible to choose the most-important variables (feature selection), eliminating the measurement of variables with the same information as the other ones. This is an aim of every technique of feature selection, which may also include the selection of transformed original variables for a better description or classification. Among these kinds of techniques we also find the search for functions that change nonnormal distributions into other ones, so that it is possible to apply the methods using models based on a normal multivariate distribution.

3 Methods and Applications

3.1 Displays

The description of large data tables by the usual univariate statistics (mean, standard deviation, range, . . .) and by histograms is still used in recent literature. Comparison between categories is made by the use of category means and ranges. Sometimes, the correlation coefficients are considered. The discussion of the extracted information can be wide-ranging and difficult to understand immediately.

However, multivariate methods of graphic description are increasingly being used to give a picture of large data tables. The main advantages are their characteristics of brevity, essentiality and ease of immediate understanding of the relevant information given by the data in the table. Eigenvector projection (EP) in two or three dimensions, nonlinear mapping (NLM) and widely used plots on the discriminant variables of the linear statistical discriminant analysis (LDA) are the methods used.

3.2 Eigenvector Projection

Eigenvector projection represents the multivariate evolution of the variable-by-variable plots. This method must be considered as the fundamental method of displaying multivariate chemical information at the beginning of or during data analysis.

Usually, the matrix of original data is column centred (by subtracting the means of the variables, column means) or column standardized (by dividing by the standard deviation of the variables) or both column centred and standardized (autoscaled). The generalized covariance matrix $\mathbf{C}_{VV} = (1/N)\,\mathbf{X}'_{VN}\mathbf{X}_{NV}$ is obtained from the centred matrix $\mathbf{X}_{NV}$; it is the same as the matrix of correlation coefficients in the case of autoscaled data. The eigenvectors of the covariance matrix are new uncorrelated variables, linear combinations of the original ones, obtained by an orthogonal rotation by a transformation matrix, the matrix of loadings, $\mathbf{L}_{VV}$.

Each column of the loading matrix stores the direction cosines of the variables of an eigenvector. After the orthogonal rotation, the matrix of the coordinates of the objects is obtained in the new system of variables, the scores,

$$\mathbf{S}_{NV} = \mathbf{X}_{NV}\mathbf{L}_{VV}\,.$$

The eigenvectors are characterized by a diagonal covariance matrix of the scores

$$\Lambda_{VV} = (1/N)\,\mathbf{S}'_{VN}\mathbf{S}_{NV}\,,$$

where all the covariances are equal to zero, so that the information given by an eigenvector is not partly copied by that given by another eigenvector, as usually happens in the case of correlated original variables. The diagonal of the matrix Λ_{VV} is the vector of the eigenvalues, and the eigenvectors are ordered with decreasing eigenvalue, so that $\lambda_j > \lambda_{j+1}$.

When we have V original variables, and no a priori reason to suppose that some variables bring more information than others (as can happen when the analytical error is very different for each variable), then each original variable, plotted on a histogram, gives (100/V) % of the total information, and a variable-by-variable plot gives (200/V) % of the total information. The plot on the first two eigenvectors gives (when the original variables have been autoscaled) $[(\lambda_1 + \lambda_2) \cdot 100/V]$ % of the total information, always greater than (200/V) % and often in the range 50%–80%.

Moreover, the last eigenvectors frequently contain useless information, noise, so that the percentage of the useful information in the plot may be greater than that calculated from the above equation. So, the plot of the scores of the first eigenvectors, to which symbols and colours add some more information (such as the category of the plotted objects), gives the greatest part of the information of the original data matrix in the simplest way for the immediate understanding of relationships, similarities, differences and dispersions of the objects. In the same way, the plot of the loadings gives the greatest part of the information on the relationships among the variables.

Eigenvectors are frequently called *factors*, when data have not been centred, and in this case the information is the distance from the origin. When data have been autoscaled and they refer to only one category, eigenvectors are called *components*, and the information is the distance from the centroid of the category. In the former case, the eigenvectors are the basis of the factor analysis, where the eigenvector transformation is followed by an orthogonal or nonorthogonal rotation in the space of the first factors. In the latter case, the name *principal component analysis* is used and the first aim is the determination of the number of significant components and the study of the loadings and scores of these components.

Because of the different names used in eigenvector analysis to indicate similar and often identical procedures, it is necessary to indicate clearly the transformation made on the original measured data before the eigenvector rotation (e.g., column centring, standardization, and also row transforms, such as the percentage of the variable in the sample, often used in chromatographic data, which removes the information given by the size of the chemical sample, and also the information given by the concentration in equal-size samples.)

Eigenvectors reduce the dimensionality of the data matrix: when the rank of the covariance matrix is $E < V$, so that $V - E$ eigenvalues vanish, or when some eigenvectors are not significant, the use of some classification methods with the scores on the first eigenvectors, instead of the original variables, can avoid singular matrices or/and noticeably speed up data analysis.

Both graphics and reduction of dimensionality, and also the use of eigenvectors in some classification methods (e.g., SIMCA) require knowledge of the number of significant components.

The inverse rotation from the eigenvector space into the space of the original variables gives exactly the original data matrix when all the eigenvectors are used:

$$\mathbf{X}_{NV} = \mathbf{S}_{NV}\mathbf{L}'_{VV} .$$

If, instead, we use $E < V$ components, the product $\mathbf{S}_{NE}\mathbf{L}'_{EV}$ differs by an "error" matrix from the original data:

$$\mathbf{X}_{NV} = \mathbf{S}_{NE}\mathbf{L}'_{EV} + \mathbf{E}_{NV} .$$

The significant components must hold the variations common to all the objects in the data matrix; the error is the individual shift from the collective behaviour. Many methods have been proposed [5, 6] to solve the problem of the significant components.

Some methods are based on the knowledge of the experimental error in the measurement of the original variables. Thus, the number of significant components is that by which the original data matrix is reproduced within the measurement error. This does not usually happen with food data, where analytical error is frequently smaller than the other individual sources of variation. The number of sources of variability in food composition is very high, and it is almost impossible that the experiment has been designed to cover all these sources of variability uniformly. So, some sources of variability appear in only one or a few objects, a minority, which behaves differently from the majority.

So, techniques used to compute the number of the significant components have to be based on the concepts of majority and prediction, as in the case of cross validation and double cross validation [6]. In recent years, this technique has been used more and more. However, it has been noticed [7] that, when two eigenvalues are very close, the uncritical use of this method can give an erroneous number of significant components, because the eigenvectors computed with a subset of the data matrix can invert their order.

When the significant eigenvectors are two or three, a few eigenvector plots of scores and loadings give all the information of the data matrix, but when the number of significant components increases, it becomes more and more difficult to obtain a definitive decision from the plots. However, most uses of the eigenvector plots in food chemistry seem to involve only two eigenvectors, and the problem of the number of significant components is ignored.

Although eigenvector plots can be used to detect anomalous objects (outliers; this is the first step of the modelling technique known by the acronym SIMCA), or to recognize the presence of some categories in an initially undifferentiated set of objects (clustering), most of the applications are in the field of identity, classification problems, as in the examples of Figs. 2–4. All three data sets will be shown again in the description of other chemometrical techniques.

The example of Fig. 3 shows the use of blind chemical analysis, where the peaks of a chromatogram do not require chemical interpretation: they are simply the fingerprint of the objects. These patterns allow the prediction of the category of an unknown object by its position in the plot, without using any classification method.

In Fig. 4, instead, we have on the same plot the scores and the loadings of the first eigenvectors, so that the mutual relationships between objects and variables can be obtained. A high value of variables 1, 2, 3, 8 causes high scores on both the components, so that the representation point falls into the top right-hand corner: the region of the category "Barolo". The discrimination between the other two categories is mainly due to variables 4, 5, 6 7. An high value of variable 4 and

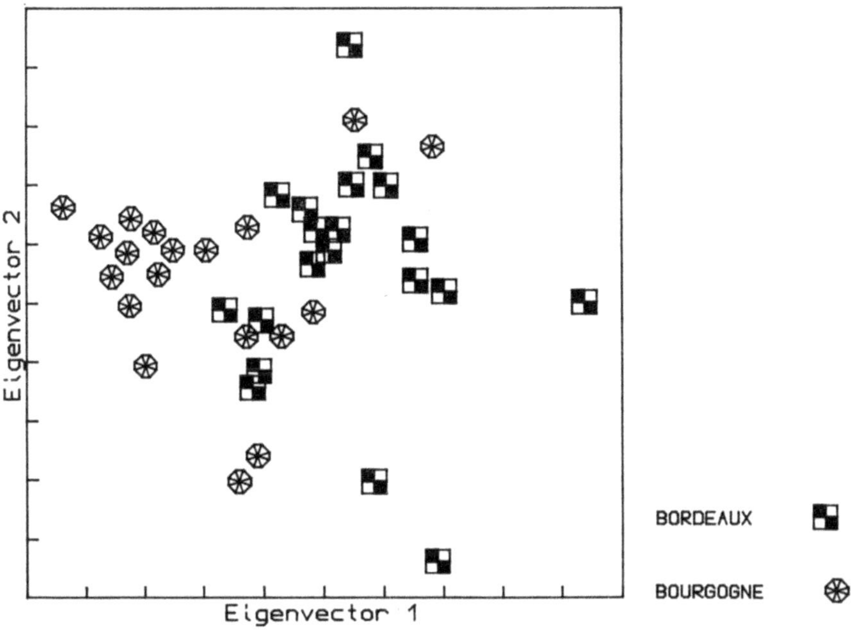

Fig. 2. Eigenvector projection of French wines, with 2 categories (Bordeaux and Bourgogne) and 20 variables (elements, organic acids, etc.). (Adapted from Ref. [8])

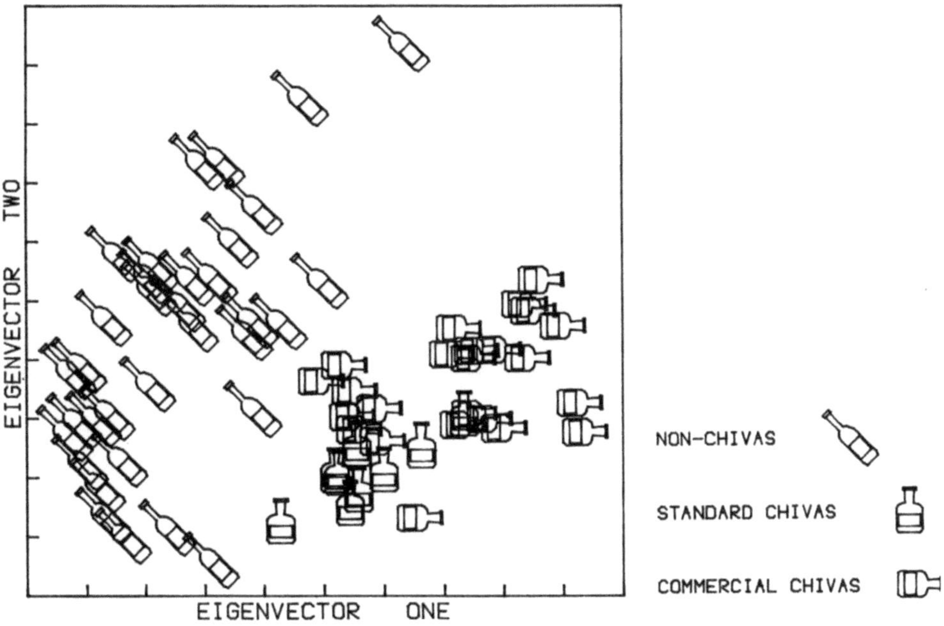

Fig. 3. Eigenvector projection of whiskies, with 3 categories (non-Chivas, standard Chivas, commercial Chivas) and 17 variables (chromatographic peaks). (Adapted from Ref. [9])

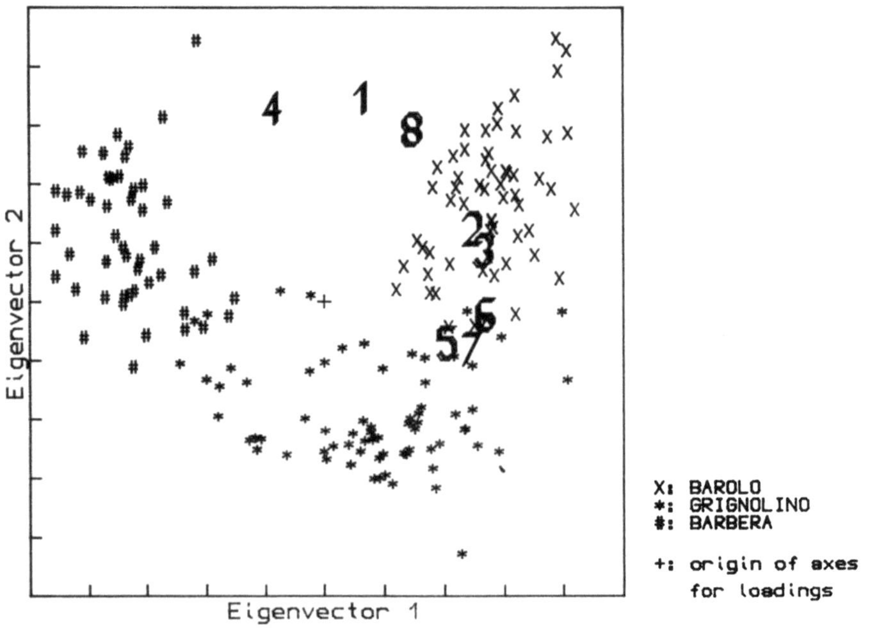

Fig. 4. Eigenvector projection of Italian red wines. 78% of the total variance retained. 3 categories (Barolo, Grignolino, Barbera); 8 variables (1: total alcohol, 2: total polyphenols, 3: flavanoids, 4: color intensity, 5: tonality, 6: O.D. 280/315 nm diluted wine, 7: O.D. 280/315 nm flavanoids, 8: proline). (Adapted from Ref. [10])

low values of variables 5, 6, 7 characterize the position of the "Barbera" samples. Variables 5, 6 and 7 give about the same information, and it is strongly negatively correlated with that given by variable 4. In the direction from variable 4 to variable 7, a set of physical variables have a great importance, whereas in the perpendicular direction the chemical variables (alcohol, proline) have the leading role.

So, the chemical direction discriminates Barolo from the other wines, the physical direction discriminates between Barbera and Grignolino. The classification ability of the selected variables is very good and probably some variables can be cancelled without noticeable loss of separation of the categories. Therefore, a small figure shows the relevant information given by a data matrix of 178 rows and 8 columns. Anyway, classification methods and feature selection methods will not modify the quality of these conclusions.

3.3 Correspondence Plots and Spectral Maps

The unified plotting of loadings and scores is a standard characteristic of two techniques: plots on the factors of correspondence analysis [11] and spectral maps [12]. These techniques are at the beginning of their use in food chemistry.

The factors of correspondence analysis are the eigenvectors of a correspondence matrix C_{VV} obtained as the covariance matrix of the original data divided by the square root of the row and column sums:

$$c_{vv'} = \sum_{i=1}^{N} \frac{x_{iv}}{\sqrt{\sum_{v=1}^{V} x_{iv} \sum_{i=1}^{N} x_{iv}}} \frac{x_{iv'}}{\sqrt{\sum_{v=1}^{V} x_{iv} \sum_{i=1}^{N} x_{iv'}}} .$$

Here the information is not the distance from the generalized mean, as in the usual EP of autoscaled data, but the distance from a mean profile, so that proportional objects ($x_{iv} = kx_{iv}$) have the same position in the plot.

Note that correspondence analysis cannot be used with centred data, and that some kind of column standardization has to be used when the range of variables is very different.

By comparing the plots of Fig. 4 and 5, we can see that in Fig. 5 the separation between categories is good (some discriminating ability is shown by factors 3 and 4 too): so the differences between the three wines are due not simply to concentration, but also to the profile of the measured quantity. Moreover, Barolo appears as a very compact category, that is, all the samples show very similar profiles, so that the wider spreading of the EP must be partially due to proportional variations.

A kind of logarithmic transform, such as ln (1 + x), is used in spectral maps within row and column centring and global standardization (division by the standard deviation around the mean of all the values of the data matrix).

Recently [13], these two techniques have been applied in food chemistry and compared with the usual EP. When some objects have characteristics very different from the

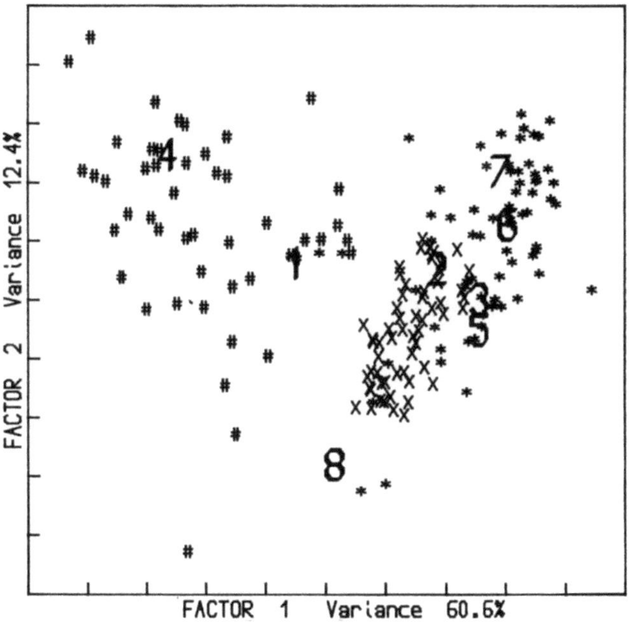

Fig. 5. Correspondence factors of Italian red wines

majority, spectral maps give a better starting representation of the experimental results, because the logarithmic transgeneration reduces the importance of these very distant objects, and the structure of the majority can be shown in the plot. Using EP, or the factors of correspondence analysis, the anomalous objects produce some directions of greater variation, which are picked out so that one among the first eigenvectors will represent the best direction to single out the anomalous objects. Then these objects may be discarded and in a further step of data analysis the remaining objects may be studied.

So, according to the aim of data analysis, one method may be selected, or another data transform may be used before the eigenvector rotation.

3.4 Nonlinear Mapping

When the significant eigenvectors are more than 2 or 3, the information cannot be easily visualized by few eigenvector plots. In these cases the use of nonlinear mapping (NLM) can give a planar representation of the objects with greater fidelity to the structure of the information in the hyperspace of the variables [14].

In NLM the objects are represented so that the euclidean distance between two objects i,i' in the representation plane $d_{ii'}$ is as close as possible to the corresponding distance $D_{ii'}$ in the space of the original variables:

$$\sum_{i=1}^{N-1} \sum_{i'=i+1}^{N} (d_{ii'} - D_{ii'})^2 = \text{minimum} .$$

This error function to be minimized includes $N(N - 1)/2$ terms; $2N - 3$ coordinates in the NLM representation plane (NLM coordinates) must be optimized (three coordinates can be fixed to avoid rotation and translation of the plot).

The search for the optimum usually starts from the coordinates in the plane of the first two eigenvectors. However, to avoid the iteration (usually done with the method of steepest descent) stopping at a relative minimum, it is advisable to repeat the search from a different starting position, such as that given by the coordinates of two original variables.

Nonlinear mapping has not been widely applied in food chemistry because:

a) the computer time becomes very great with an increasing number of objects;

b) the plot coordinates are not directly linked to the original variables, so that it is impossible to evaluate the relative importance of the original variables as regards the separation of the clusters of objects or the presence of outliers;

c) when one or more objects are added to the data matrix, the NLM must be completely repeated to display the new objects.

An example of the use of NLM is shown in Fig. 6. Nonlinear mapping was also used in the representation of the amino acid spectrum of French red wines [16]: 110 objects (Bordeaux, Beaujolais and non-Beaujolais Bourgogne wines) characterized by 20 amino acids were represented.

To reduce computing time, a two-step method has been suggested [17], the simplified nonlinear mapping (SNLM), and it was first used to present samples of olive oil of several Italian regions [18]. In the first step, the NLM coordinates of a

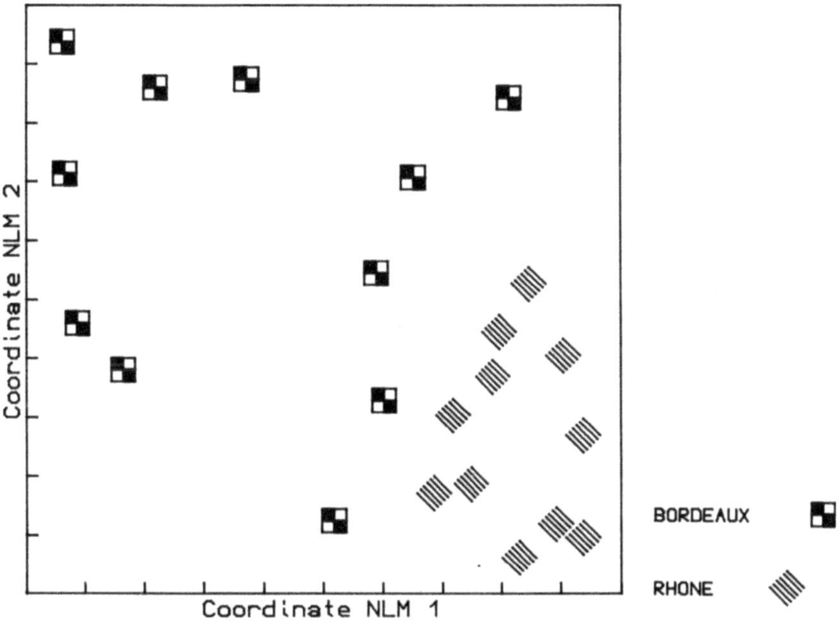

Fig. 6. Nonlinear mapping of Bordeaux and Rhone wines. (Adapted from Ref. [15])

reduced number of points are computed in the hyperspace of the variables. These points (base points) can coincide with some selected objects or they can be the centroids of the categories or else arbitrary points. The NLM of the base points requires a short computing time.

In the second step, all the objects of the data matrix are plotted in the plane of the SNLM coordinates, one at a time, so that in the plot the distances between objects and base points are as close as possible to those in the original variable hyperspace. So, instead of one optimization problem with 2N — 3 coordinates to be optimized, we have N optimizations of 2 variables.

The starting point of each object is computed by a geometrical consideration: in the representation plane there are two specular points equally distant from the two closest base points, as in the variable hyperspace. One of these two points is closer to the third-nearest base point, and this one is chosen as the starting point for the optimization procedure.

Figure 7 is the SNLM of the data set whose eigenvector projection is shown in Fig. 4. The separation between the three categories is worse than that given by EP, but this is not necessarily a negative characteristic, because in the SNLM plot all the information is represented, while EP discards the information of the minor eigenvectors. But there is a negative side: in the plot, a nonsignificant split of each category into two apparent subcategories appears. This splitting is frequently observed with few base points 3 or 5, and it can be avoided [10] by using a correction factor for the distances in the space of original variables, if the number of base points is less than the number of variables. This factor is the square root of the

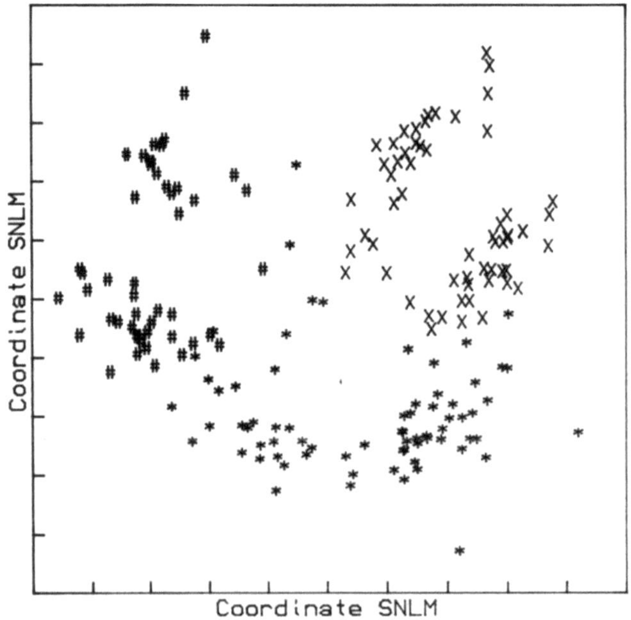

Fig. 7. Simplified nonlinear mapping of Italian red wines (category centroids as basepoints). 8 variables (see Fig. 4)

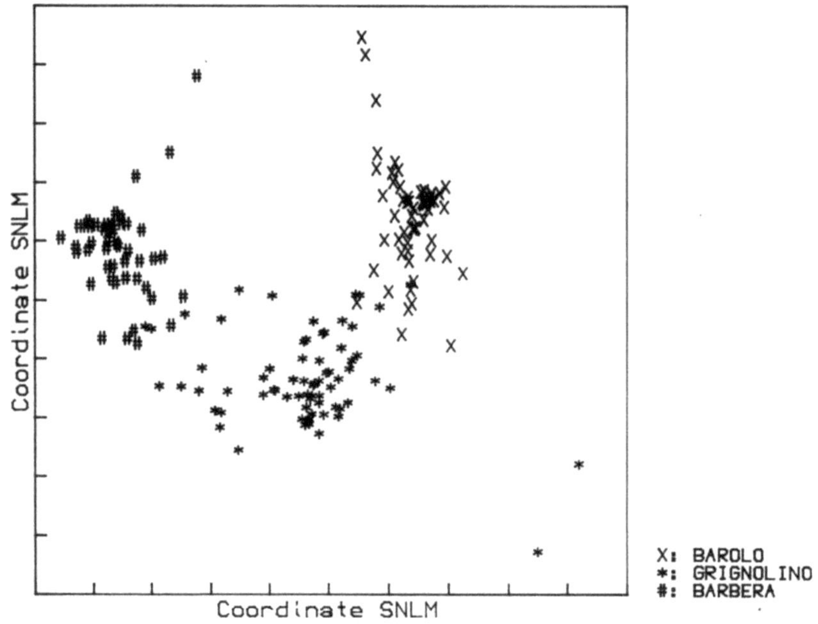

Fig. 8. Simplified nonlinear mapping of Italian red wines; with correction factor. 8 variables (as in Fig. 4)

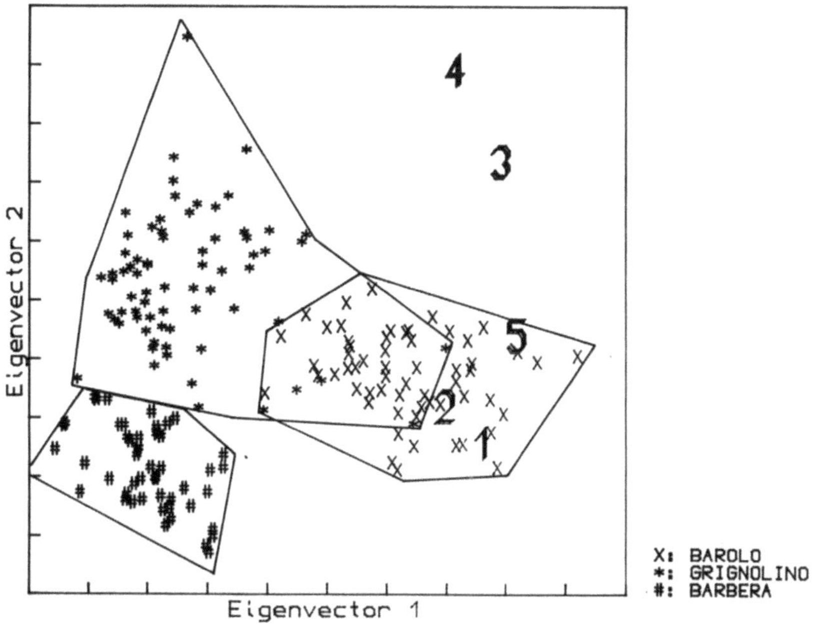

Fig. 9. Eigenvector projection of Italian red wines. 5 variables (1: flavanoids; 2: total alcohol; 3: tonality; 4: Mg; 5: proline)

ratio between the number of base points and that of the variables. Figure 8 shows how the use of the correction factor produces a plot without subcategories.

The previous figures show that the use of NLM does not give a real advantage when the useful discriminating information is contained in the plane of the first two eigenvectors.

In Figs. 9, 10, the same data set of 178 wines was studied using five variables only (it has been shown [19] that these five selected variables give the maximum predictive ability, about 100%, in the classification of the three wines). Here the eigenvector projection (Fig. 9) shows heavy overlapping between two categories, whereas the SNLM indicates very little overlapping, so the SNLM plot produces a better representation of the discriminating information given by the five variables.

3.5 Plot on the Discriminant Functions

Because of the difficulty of selecting the right transforms before EP, the long computing time of NLM, and because the programs for NLM or SNLM are not generally available, the most-used representation method for classification problems is the discriminant function plot of the linear statistical discriminant analysis (LDA). Well-known packages contain this method, which does not require preliminary treatments of the variables or a long computing time. Figures 11–19 show examples of the use of this kind of representation.

In Sect. 4.1 we will discuss the method of linear statistical discriminant analysis. Here, however, some comments are given in advance.

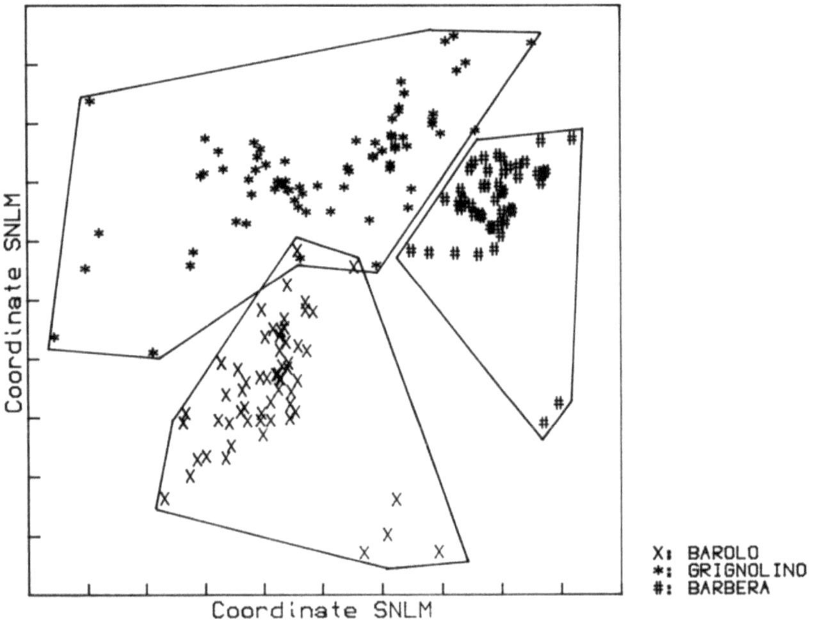

Fig. 10. Simplified nonlinear mapping of Italian red wines. 5 variables (as in Fig. 9)

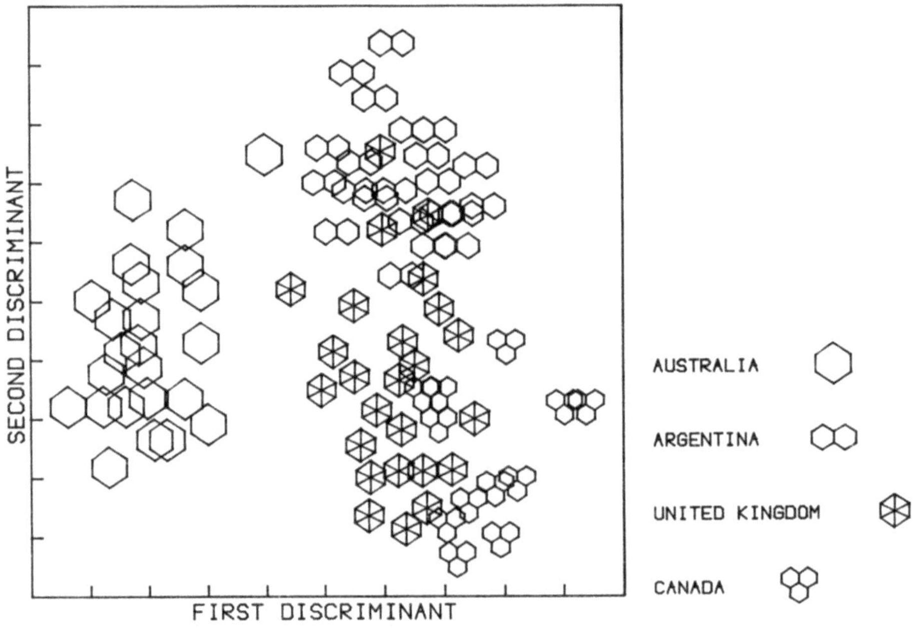

Fig. 11. Plot on the discriminant functions of LDA. 4 categories of honey (Australia, Argentina, United Kingdom, Canada). 17 variables (amino acids). (Adapted from Ref. [20])

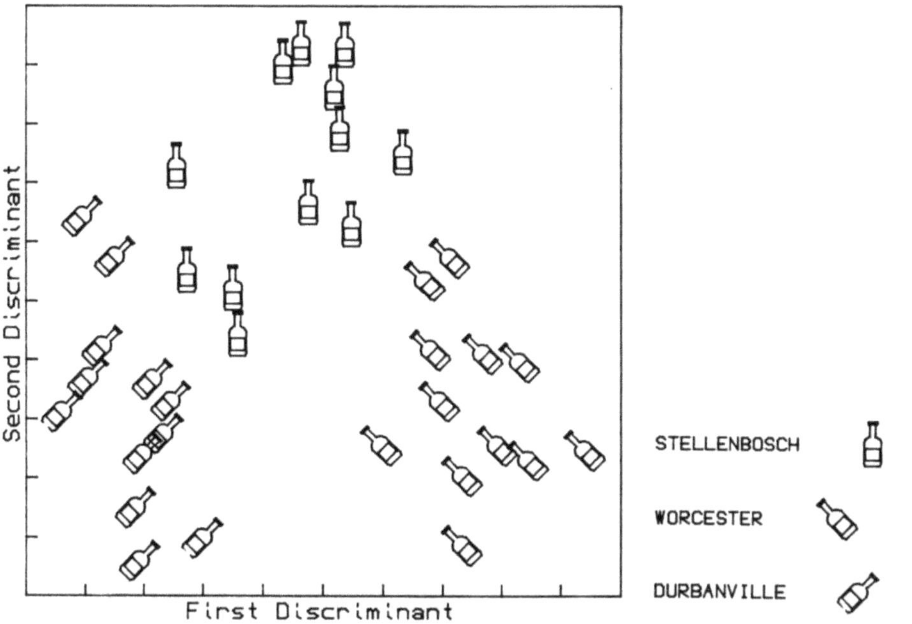

Fig. 12. Plot on the discriminant functions of LDA. 3 categories of wine (Pinotage wines from Stellenbosch, Worcester, Durbanville). 16 variables (esters, alcohols, acids). (Adapted from Ref. [21])

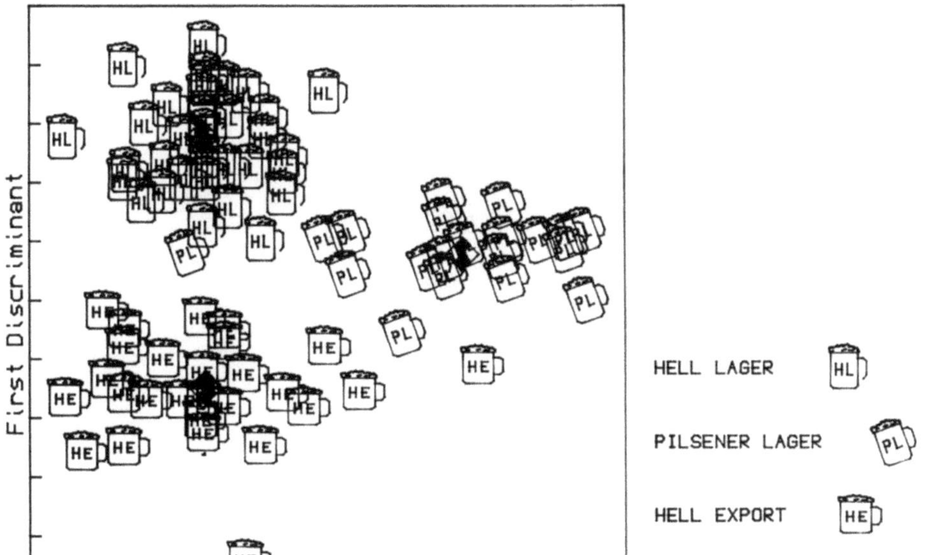

Fig. 13. Plot on the discriminant functions of LDA. 3 categories of beer (hell lager, pilsener lager, hell export). 8 variables. (Adapted from Ref. [22])

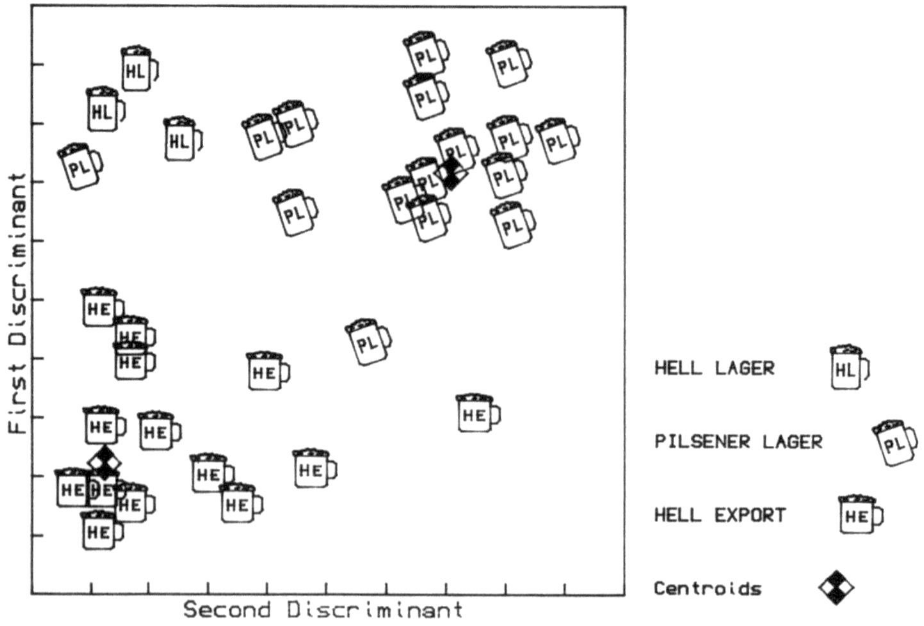

Fig. 14. Enlargement of the boundary area of Fig. 13

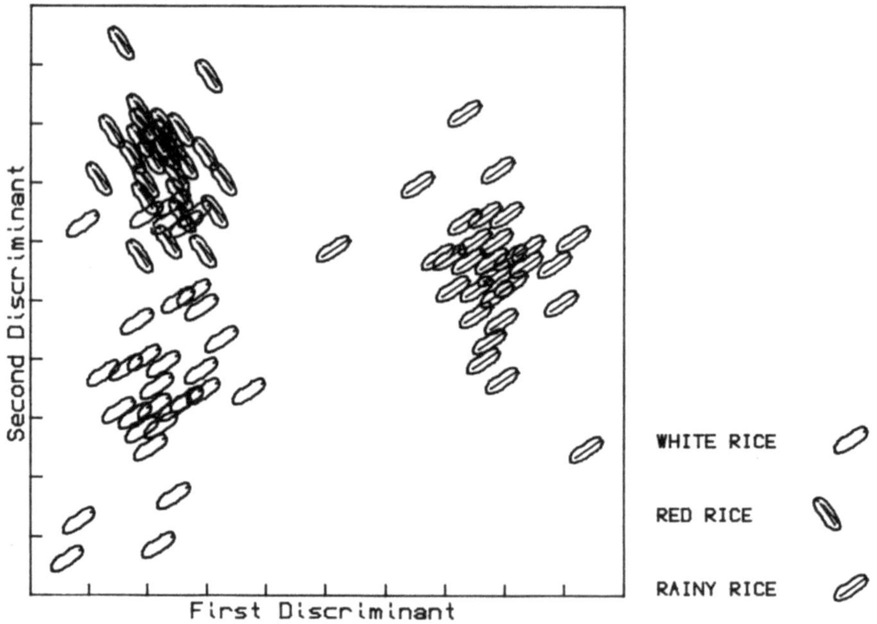

Fig. 15. Plot on the discriminant functions of LDA. 3 categories (white rice, red rice, rainy rice). 12 variables (total lipids, rough proteins, fatty acids). (Adapted from Ref. [23])

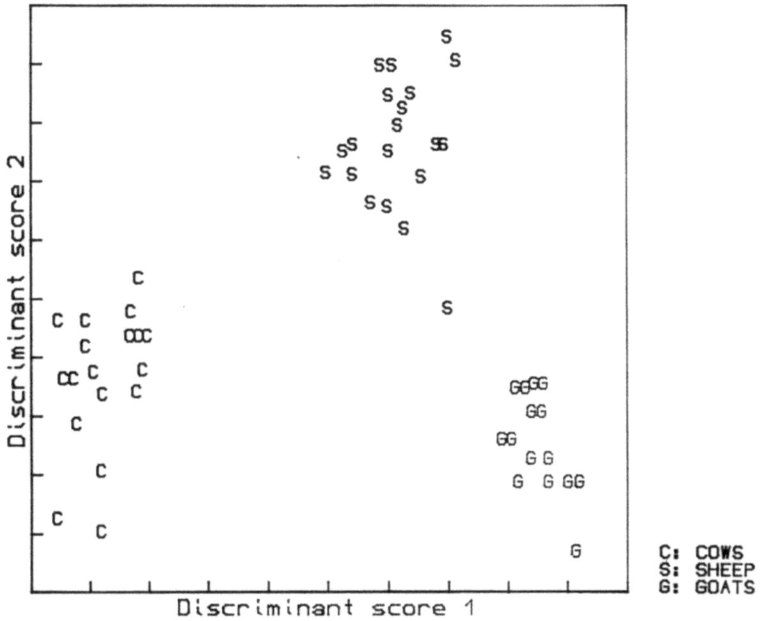

Fig. 16. Plot on the discriminant functions of LDA. 3 categories (milk of cows, sheep, goats). 15 variables (fatty acids). (Adapted from Ref. [24)])

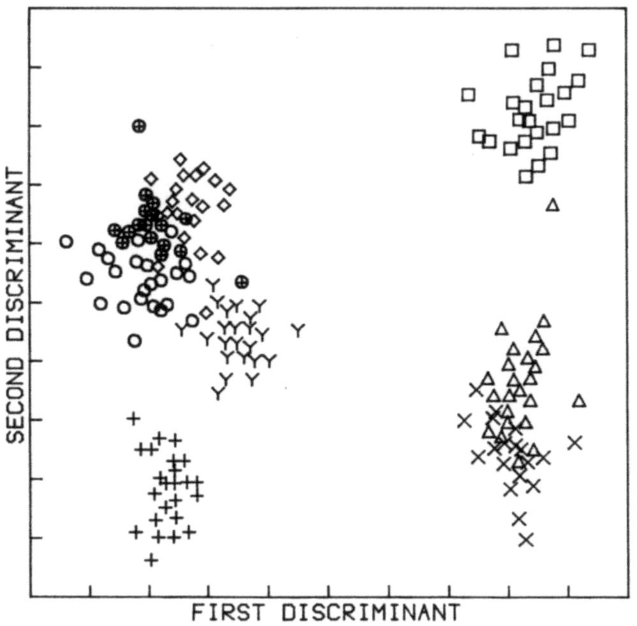

Fig. 17. Plot on the discriminant functions of LDA. 8 categories (brands of soy sauce). 25 variables (gas chromatographic peaks). (Adapted from Ref. [25)])

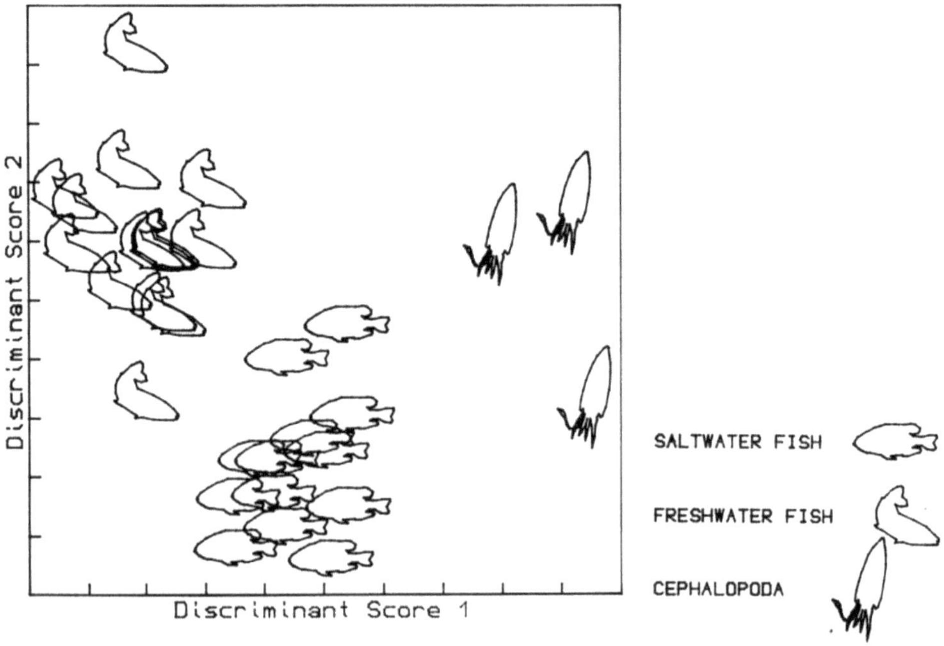

Fig. 18. Plot on the discriminant functions of LDA. 3 categories (saltwater fish, freshwater fish, cephalopoda). 10 variables (fatty acids). (Adapted from Ref. [25], after correction)

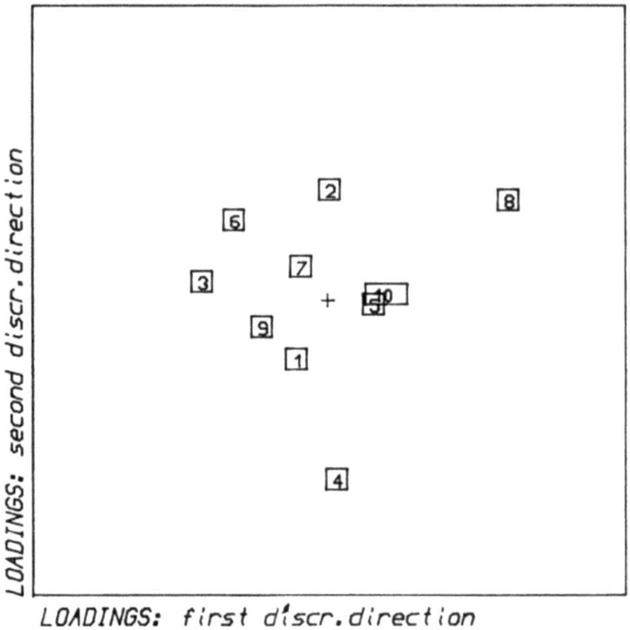

Fig. 19. Loadings on the canonical functions (data as for Fig. 18). 10 variables (fatty acids)

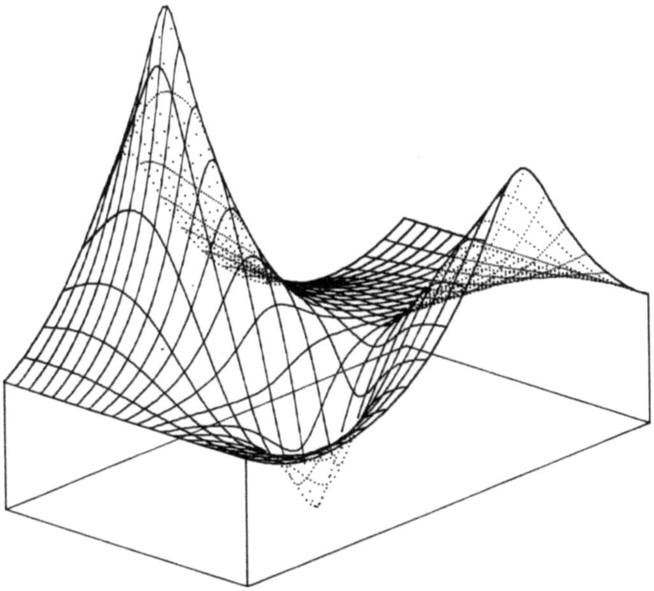

Fig. 20. Ratio of the interclass to intraclass variance as a function of zenith and azimuthal angles in a simulated case with 3 variables and 3 categories

Whereas EP and NLM may be used with categorized or noncategorized data, LDA searches for directions (in the space of original variables) with maximum separation between categories. These directions are the eigenvectors of a nonsymmetric matrix obtained by the intercategory covariance matrix (or the intercentroid covariance matrix or the generalized covariance matrix) premultiplied by the inverse of the intracategory pooled covariance matrix [27-29]. Because of the lack of symmetry in this matrix, the transformation from the original variables to the discriminant functions is a nonorthogonal transformation, with a consequent distortion of the data space. Whereas the first eigenvector of a covariance matrix is the direction of maximum variance (independent of categories), the first discriminant direction is that of maximum ratio between intercategory and intracategory variances (Fig. 20).

This ratio can be significantly changed by adding or removing a few objects, especially if the number of objects is not many times that of the variables, or if one category is formed by only a few objects. So, some care is needed in using this kind of plot. It is worth redoing the plot after the removal of some objects and observing the changes of the loadings of variables on the discriminating functions.

3.6 Other Methods

Colour and three-dimensional graphics [4] improve the display of information in food chemistry; they may be applied to original variables, to eigenvectors, to NLM coordinates and to discriminant functions. Actually, however, this kind of plot cannot be easily used in chemical journals and books, due to both high costs and technical difficulties.

Other methods seem to require some attention: Van der Voet and Doornbos have used two-dimensional pictures of human faces [30] to represent samples of Bourgogne and Bordeaux wines, where the original variables determine the anthropometrical variables of the faces.

The method of audio representation of multivariate analytical data [31] has not been used, to our knowledge, in the representation of food data. However, the large number of properties of sound (pitch, loudness, damping, direction, duration, rest) seems suitable for the recognition of complex information, at least as an alternative.

So, we can imagine a future of the representation techniques in which three-dimensional coloured figures with a pleasant (or otherwise) accompaniment are used to represent food analytical data with clear relationships to the food quality and origin. We leave to the reader's imagination the evaluation of the possibilities offered by the other human senses in the field of data representation.

4 Classification and Modelling Methods

In recent years, new methods have been introduced into chemistry for classification problems, and they have often been applied to food analytical data. The statistical linear discriminant analysis is still the most widely used method, as was noted in the previous section.

4.1 Linear Statistical Discriminant Analysis

The assumption of multivariate normal distribution underlying this method gives for the conditional (a posteriori) probability density of the category g [32]

$$p(\mathbf{x}/g) = \frac{1}{(2\pi)^{V/2}\,|C_g|^{1/2}} \exp\left[-\frac{1}{2}(\mathbf{x}-\bar{\mathbf{x}}_g)\,\mathbf{C}_g^{-1}(\mathbf{x}-\bar{\mathbf{x}}_g)\right],$$

where $\bar{x}_g$ and c_g are the mean and the covariance matrix of each class. The term $(\mathbf{x}-\bar{\mathbf{x}}_g)'\,\mathbf{C}_g^{-1}(\mathbf{x}-\bar{\mathbf{x}}_g)$ is known as the *Mahalanobis distance*.

With the additional assumptions of equal a priori probability p(g) for each class, and that the covariance matrices are the same for each class, P, the logarithm of the product p(g) p(**x**/g) becomes $\ln[p(g)\,p(x/g)] = \text{const.}\left(-\frac{1}{2}(\mathbf{x}-\bar{\mathbf{x}}_g)'\,\mathbf{P}^{-1}(\mathbf{x}-\bar{\mathbf{x}}_g)\right)$.

The higher it is, the higher is the so-called *discriminant score* (also known as the *discriminating function*)

$$S_g(\mathbf{x}) = \mathbf{x}'\mathbf{P}^{-1}\bar{\mathbf{x}}_g - \frac{1}{2}\bar{\mathbf{x}}_g'\mathbf{P}^{-1}\mathbf{x}_g,$$

where the terms independent of the category have been eliminated.

The lines of equal probability (and of equal discriminant score) appear as hyperellipsoids in the space of the variables, equal for each class (see Fig. 21); the locus, where

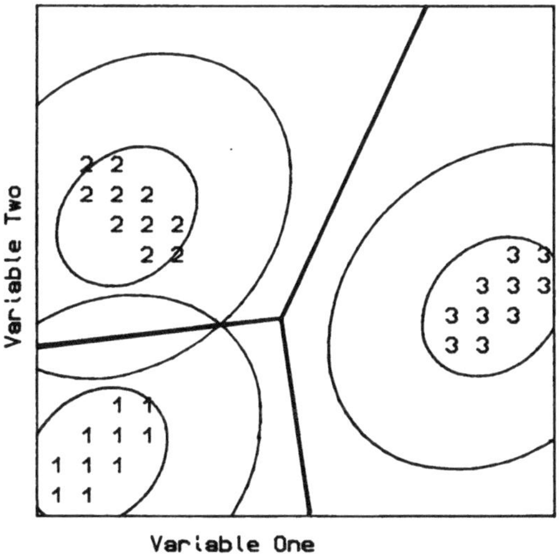

Fig. 21. Equiprobability ellipses and discriminant lines for statistical linear discriminant analysis (bivariate case)

the discriminant scores for two classes, 1 and 2, are the same, is the line

$$S_1(\mathbf{x}) - S_2(\mathbf{x}) = 0 = \mathbf{x}'\mathbf{P}^{-1}(\bar{\mathbf{x}}_1 - \bar{\mathbf{x}}_2) - \frac{1}{2}\bar{\mathbf{x}}_1'\mathbf{P}^{-1}\bar{\mathbf{x}}_1 + \frac{1}{2}\bar{\mathbf{x}}_2'\mathbf{P}^{-1}\bar{\mathbf{x}}_2 \,,$$

and this line is the linear statistical delimiter between the two classes.

In the space of the canonical discriminant functions, the pooled covariance matrix is a constant multiplied by the identity matrix, the equiprobability hyperellipsoids become hyperspheres, and the linear statistical delimiter is a line midway between the centroids of two categories. This kind of plot has been used many times by Scarponi and co-workers [33–35] in the classification of white wines.

Some computer programs draw the equiprobability circles at a selected confidence level. So it is possible to evaluate how close an unknown object is to a category: in this way LDA approaches the modelling methods. Gilbert et al. [20] used 95% confidence circles to show the separation between honeys from the United Kingdom, Australia, Argentina and Canada.

The observation of the linear delimiters and of confidence circles (or ellipsoids) can often indicate that the assumptions for LDA are not completely satisfied. For example, we can see in Fig. 22 that (a) honeys from Canada show a greater variance in the direction of the second discriminant function (really, the number of objects in this category is too small to give probatory information), (b) one sample from the United Kingdom falls very far from the centroid of this category, and the other samples are about on a line in a direction close to that of the first variate. Only the structure of the category "Argentinian honey" appears to be well described by the assumptions of linear statistical discriminant analysis.

Wenker et al. [36] used dispersion polygons to show the differences in the dispersion of some classes of French brandies. The number of classification errors on the basis of these dispersion polygons appears to be very small in comparison with that based on the discriminating scores (i.e., when the canonical discriminating functions are used, and the classification made on the basis of the distance from category centroids).

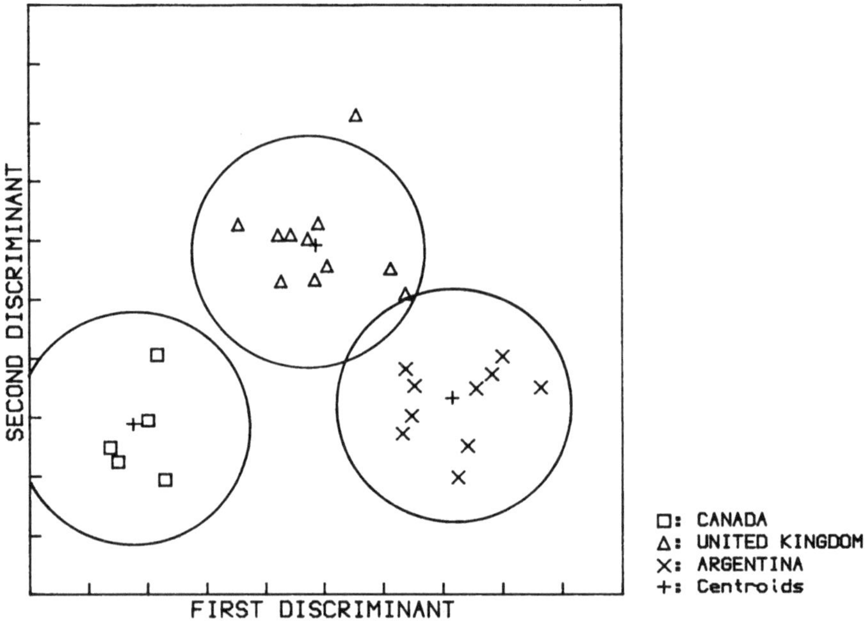

Fig. 22. Confidence circles in the space of the canonical functions of LDA. 3 categories (honeys from Canada, United Kingdom, Argentina). 17 variables (amino acids). (Adapted from Ref. [20])

These efforts to improve the classification ability by correction of the original LDA model with the use of graphical means that search for a better discriminant line are the prelude to the use of classification methods with separate class models: the *bayesian analysis*.

The terms *classification ability* and *classification error* used above refer to a procedure in which we have N objects of G categories, and all objects are used to compute class means and a pooled intraclass covariance matrix. The objects are then classified according to their highest discriminant score. This procedure is that commonly used. However, the classification ability (percentage of correctly classified objects) is an overestimate of the real utility of information, which must be considered as the ability to classify correctly unknown samples: predictive ability.

Two methods are used to evaluate the predictive ability for LDA and for all other classification techniques. One method [37] consists of dividing the objects of the whole data set into two subsets, the training and the prediction or evaluation set. The objects of the training set are used to obtain the covariance matrix and the discriminant scores. Then, the objects of the training set are classified, so obtaining the apparent error rate and the classification ability, and the objects of the evaluation set are classified to obtain the actual error rate and the predictive ability. The subdivision into the training and prediction sets can be randomly repeated many times, and with different percentages of the objects in the two sets, to obtain a better estimate of the predictive ability.

The second method, the leave-one-at-a-time or jackknife procedure, repeats the whole LDA procedure as many times as there are objects, and each time one object alone is the evaluation set.

The two methods have been compared by Moret et al. [35] in the classification of three white wines: they give about the same predictive ability, lower, of course, than the classification ability. The difference between classification and predictive ability becomes greatest when the number of objects is not much greater than the number of variables, or when there are very few objects in a category. A very low predicitive ability and a high classification ability reveal a bad experimental design.

Baldi et al. [38] used LDA in the classification of wines of the same cultivar (Barbera) and different geographical origin. There were 13 samples (6 in the first two categories, 1 in the third) and 7 variables were selected out of 20 measured quantities. The obtained 92% classification ability is meaningless, especially with regard to the third category.

This second fault of LDA, that is, the impossibility of obtaining significant information when the ratio between the numbers of objects and variables is low, leads to the use of SIMCA, the modelling method to be chosen in this case.

4.2 Bayesian Analysis

This classification method has not been widely used in food chemistry, probably because the related computer programs are not as widespread as that of LDA. However, in recent years, some classification problems have been analysed by this method, and the results show a predictive ability generally better than that obtained by LDA, so a wider use of bayesian analysis (BA) appears desirable.

The main characteristics of BA are [32]:

a) Each category has a separate model, given by the centroid of the category and the covariance matrix of the category. (In the BA performed by program BACLASS of the package ARTHUR, the class model is given by the marginal histograms, smoothed by a suitable function.)
b) The conditional probability density is computed from the multivariate normal distribution. (In ARTHUR-BACLASS, the conditional probability density is computed as the product of the marginal (univariate) probability densities, obtained from the histograms, or from the mean and standard deviation of each variable, under the assumption of independent variables.)
c) To obtain the decision function, the a priori probability of each class is taken into account, within a loss function given by a matrix L. Each element l_{ij} of this matrix L specifies the loss associated with the classification of an object of category i into category j. Usually, $l_{ij} = 0$ when $i = j$, and $l_{ij} = 1$ when $i \neq j$ (loss 0 for correct classification, loss 1 for every misclassification: normalized loss).

   The decision function

$$f_g(\mathbf{x}) = \sum_{j=1}^{G} l_{jg} p(\mathbf{x}/j)\, p(j)$$

(object x is assigned to the class g with the minimum value of the decision function) becomes, in the case of normalized loss

$$d_g(\mathbf{x}) = p(\mathbf{x}/g)\, p(g)$$

(the object x is assigned to the class g with the maximum value of the decision function).

The posterior probability, i.e., the probability that the object belongs to class g, is given, in this case, by

$$p(g/\mathbf{x}) = d_g(\mathbf{x}) \Big/ \sum_{g=1}^{G} d_g(\mathbf{x}) .$$

To our knowledge, BA has always been used with the normalized loss function, and with $p(g) = 1/G$ independent of the category. However, these two elements of the bayesian decision can be of greatest importance in the practical use of the classification technique with unknown objects. The loss from misclassification is meaningful depending on the values (commercial, nutritional, cost of further analysis, . . .) of each category. So, if we have a hypothetical classification problem with two high-quality and two low-quality categories, the loss associated with the misclassification of a high-quality object into the other high-quality category is lower than its misclassification into one of the two low-quality categories. Because of this, we believe that the development of a decision function to be used in the usual practice of food laboratories should be preceded by an exhaustive study of the consequences of misclassification.

d) The squared Mahalanobis distance in p(x/g) when the multivariate normal distribution is applied is useful in obtaining the hyperellipsoids of equal probability that fix the boundaries of the multivariate confidence interval of each category. The confidence hyperellipsoid is the model of the category to be used when BA is applied as a modelling technique.

In a modelling technique, attention is paid to the category space (in the case of BA, the confidence hyperellipsoid of the category), not to the discriminant delimiter lines as in the classic classification methods. Each category is studied separately and objects (both of the studied category or of other categories, both in the training or in the evaluation set) are classified as objects fitting the category model (if they fall within the category space), or as objects outside the category model.

Then the classification is made for each object, taking into account that an object can fit the model of just one category or more than one category (when two or more category spaces overlap, at least partially) or it can be outside every category space, an outlier.

The squared Mahalanobis distance has also been used as a decision function in BA [39]:

$$d_g(\mathbf{x}) = (\mathbf{x} - \bar{\mathbf{x}}_g)' \, \mathbf{C}_g^{-1}(\mathbf{x} - \bar{\mathbf{x}}_g)$$

(the object x is assigned to the class g with the minimum value of the Mahalanobis distance). The use of this decision function is not advisable when the determinants of the covariance matrices of two categories are very different and the category spaces overlap. In the univariate example of Fig. 23, it can be seen that the x-interval within which an object is classified into category b by the Mahalanobis distance (MD) rule is smaller than that obtained by the probability density (PD) function. The interval vanishes when the two distributions have the same centroid, so that the information cannot be used for classification purposes. However, this is without importance when BA is used as a modelling technique and no classification decision is taken when an object fits the model of two categories.

Moreover, because the Mahalanobis distance is a chi-square function, as is the SIMCA distance used to define the class space in the SIMCA method (Sect. 4.3), it is possible to use Coomans diagrams (Sect. 4.3) both to visualize the results of modelling and classification (distance from two category centroids) and to compare two different methods (Mahalanobis distance from the centroids versus SIMCA distance).

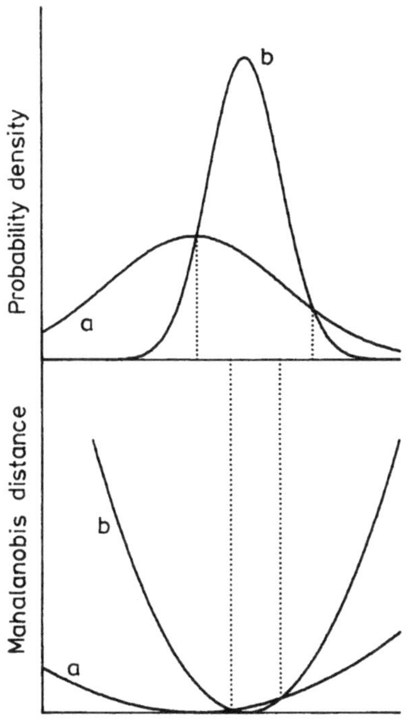

Fig. 23. Comparison of classification by probability density functions and by Mahalanobis distance. Univariate case. Values in the range between the dotted lines are classified into class b

Both PD- and MD-based BA should be applied only after testing the normal distribution of objects [40]; however, good results have been obtained, as regards prediction ability, with skewed distributions too, proving the high robustness of this method.

The Bayesian analysis of BACLASS (a program of ARTHUR), where the decision function is obtained from the product of the marginal PDs computed by the smoothed (symmetrical or skewed) histograms, may apparently be used with skewed distributions, without preliminary transformations of the original variables.

Because of the assumption of independent variables, the space class is computed as a hyperellipsoid, whose axes are parallel to the variables and whose volume is much greater than that of the corresponding hyperellipsoid of the classical BA. The decorrelation of variables by subtraction (program SELECT of ARTHUR) or the use of the scores of eigenvectors instead of original variables cannot give any improvement. In practice, the decorrelation is made on the whole data set, and the correlation is cancelled in the generalized covariance matrix, not in the category covariance matrices.

The category correlations can be cancelled only when all the objects of the training set are in the same category, and the method is used as a class modelling technique. However, the bayesian analysis in ARTHUR-BACLASS has been compared with the usual BA in classification problems about wines and olive oils [10] and about the same classification and prediction abilities were observed for both methods.

A wider comparative use of these bayesian methods (PD-BA, MD-BA, BACLASS) seems useful, because the relative performances would probably change according to the skewness and correlation of variables, and the overlap of categories.

Like linear discriminant analysis, bayesian analysis can be used only when the number of objects is higher than that of variables. In addition, BA requires this condition in each category, and an objects/variables ratio greater than two in each category of the training set is highly desirable. Moreover, also with a high objects/variables ratio, the rank of a category covariance matrix may be less than the number of variables (due to linear relationships between variables, such as that given by percentage row transformations often used, e.g., with chromatographic data).

In this case, the multicollinearity problem can be avoided by feature selection (e.g., by a decorrelation procedure, subtraction or eigenvector projection) followed by BA in the space of selected features.

In summary, BA has been used in the space of principal eigenvectors in problems about oils and wines, and the plot of the first two eigenvectors has been used to display the confidence ellipsoids (class space) and their changes after outlier deletion: this is the procedure to obtain an improved class model.

4.3 SIMCA

Both LDA and BA have as the centroid of a category the point with the highest probability density, and the greater the distance of an object from it, the lower the probability that the object will be in that category. The model of BA is a point, and each object is described by it and by an error vector

$$\mathbf{x}_{ig} = \bar{\mathbf{x}}_g + \boldsymbol{\varepsilon}_i \,.$$

The class space is the space around the centroid in which errors have some permitted values (according to the underlying selected distribution).

SIMCA (Soft Independent Modelling of Class Analogy) [41–43], is the first modelling and classification technique in which the model may be linear, piece-wise linear, planar or a multidimensional figure. Most of the theory of other recent modelling techniques derives from SIMCA. Originally, it was developed and used with many variables and few objects, where LDA and BA cannot be used except after selecting a small number of features. However, an increasing number of people use SIMCA when BA and LDA could also be applied. We believe that this is not purely a wish to apply a new method, but is a consequence of some special characteristics of the method itself.

a) In SIMCA the (known or unknown) factors that form the vector x (i.e., in many cases, the composition of a sample) are divided into inner and outer factors. The number of inner factors fixes the model space (IMS, inner model space [44]) where the variations dependent on structure, the correlations, the possibility of interpretation and often of identification of the underlying factors are collected. The outer space, instead, collects random errors, uncorrelated variations and the effect of factors that concern only a minority of the objects.

   These concepts are surely of great importance in the identity problems of food chemistry. Here we collect, under the name of a class, objects whose chemical composition depends on several factors (treatments, ageing, etc.).

   In the following example, the measured chemical quantities are fixed by four continuous factors: age, treatment A, treatment B and analytical error. Treatment B has been applied to only one sample of the class. Age and treatment A produce variations in composition that cannot be interpreted as deviations from the model: they are the inner factors. On the other hand, treatment B cannot be identified as an inner factor because of the lack of information, and its effects fall, with the analytical error, in the outer space. Besides, if the effects of treatment B are noticeably greater than those of the analytical error, the B-treated object can be identified as an outlier, as it really is.
b) Although it is not necessary, SIMCA obtains the principal component of each category by Nipals algorithm, then it is possible to handle data matrices with some data missing and to apply the double cross validation to obtain the number of significant components: two very nice characteristics.
c) The inner model space can collect the greatest part of the skewness of the distributions, so that the variations in the outer space are often normally distributed. When this does not happen, the abnormal variations identify a few outliers, which are deleted to obtain the improved model.

These and other favourable characteristics explain the increased application of SIMCA in food classification problems, and the development of slightly modified methods or new techniques that try to improve SIMCA, retaining its main characteristics.

The class model of SIMCA is a line, a plane, etc., according to the number of significant components. By splitting a category into two or more categories, a piecewise linear model may be obtained. The boundaries of the model are obtained by the range of the scores (of objects in the training set) of the components. This "normal" range is an underestimate of the true range when the number of objects is low (see Fig. 24). So, the final range of each component is obtained by increasing and decreasing the upper and lower limits, respectively, by one standard deviation of the distribution on the component. Because of the uncertainty in the range, the deviations in the direction of a component outside the final range (as may happen for objects that have not been previously used to obtain the class model) are weighted by the inverse of the standard deviation of the scores on the component.

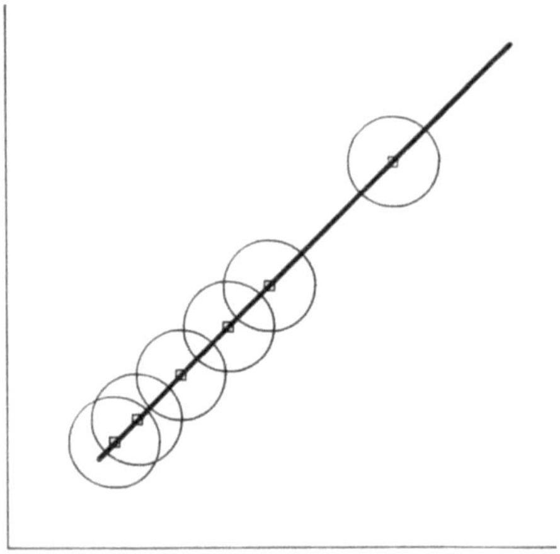

Fig. 24. Random sample of a few (6) objects from a rectangular distribution (one component, two variables) plus random error (circles). The experimental range of the component is an underestimate of the true range

The usual enlarged model has been used in some applications of SIMCA to food identity problems [8,45–49]. Sometimes, the results compared with those deriving from other techniques were not as good as expected. However, it was noticed [10] that the enlarged model can hardly be applied when the number of objects is large. In this case the normal range may be an overestimate of the true range, because of the contribution of random errors in the direction of the component (see Fig. 25). Consequently, a "reduced" SIMCA model has been proposed, which is obtained by a both-sides contraction of the normal range. As a result, classification and predictive abilities obtained in the classification of Portuguese olive oils [10] and some Italian wines [50] were about the same of those of bayesian analysis.

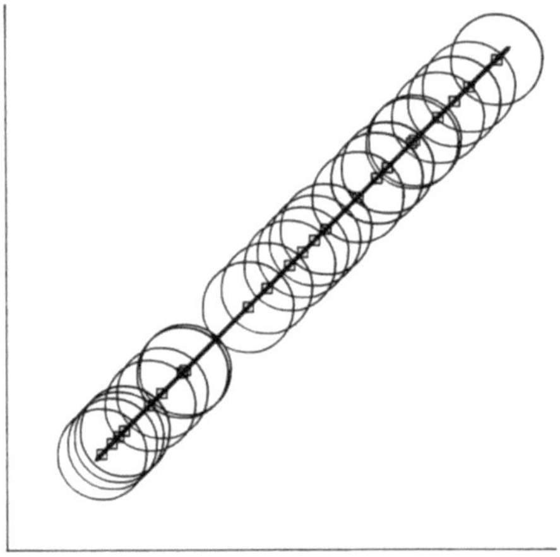

Fig. 25. Random sample of many (25) objects from a rectangular distribution (one component, two variables) plus random error (circles). The experimental range of the component is an overestimate of the true range

Objects do not fall exactly into the inner model space, and a residual error on each variable can be computed. These residuals are uncorrelated variables, because each significant correlation is retained in the linear model. So, the variance of residuals is a chi-square variable, the SIMCA distance, and, multiplied by a suitable coefficient obtained from the F distribution, it fixes the boundary of the class space around the model, called the SIMCA box, that corresponds to the confidence hyperellipsoid of the bayesian method. Objects, both those used and those not used to obtain the

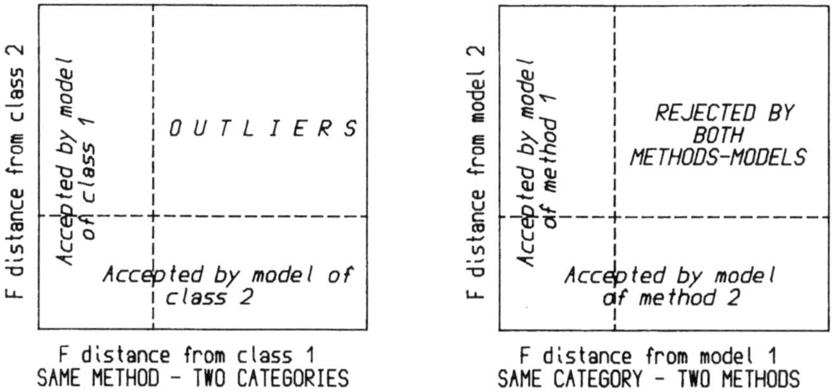

Fig. 26. Coomans diagrams

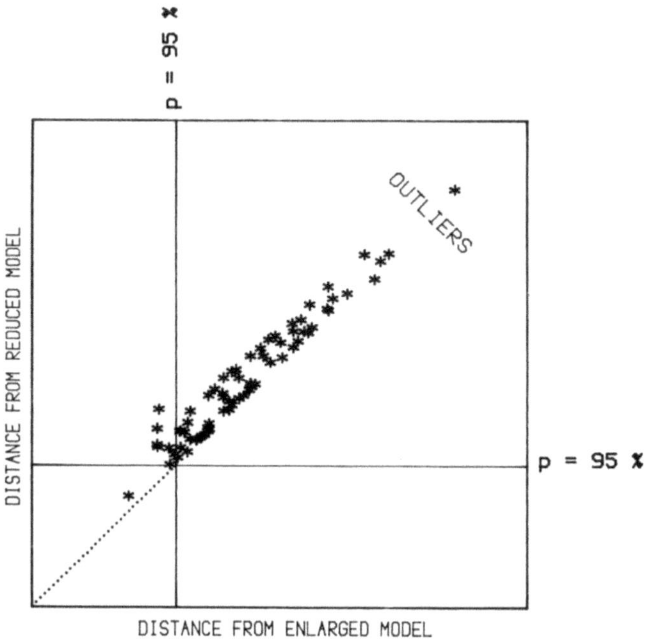

Fig. 27. Coomans diagram for SIMCA usual (enlarged) and reduced models. The fitting of Grignolino samples to the two models of the Barolo class [10] is shown

class model, are able to be classified into the class or outside the class according to the standard deviation of their residuals.

The SIMCA distances from two class models (or from the two models of the same category obtained by different methods) are reported in Coomans diagrams [51] (Fig. 26) to show the results of modelling-classification analysis.

Figure 27 shows an application of a Coomans diagram to a comparison between the usual (enlarged) and reduced SIMCA models, showing that the reduced model is less easy to penetrate than the usual model.

4.4 KNN and Potential Function Methods

Among the nonparametric techniques of pattern recognition, the linear learning machines [52] have been only seldom used in food data analysis [8,48], and it seems that this method is becoming obsolete.

The rule of the K nearest objects, KNN, has been used in classification problems, in connection and comparison with other methods. Usually KNN requires a preliminary standardization and, when the number of objects is large, the computing time becomes very long. So, it appears to be useful in confirmatory/exploratory analysis (to give information about the environment of objects) or when other classification methods fail. This can happen when the distribution of objects is very far from linear, so that the space of one category can penetrate into that of another, as in the two-dimensional example shown in Fig. 28, where the category spaces, computed by bayesian analysis or SIMCA, widely overlap.

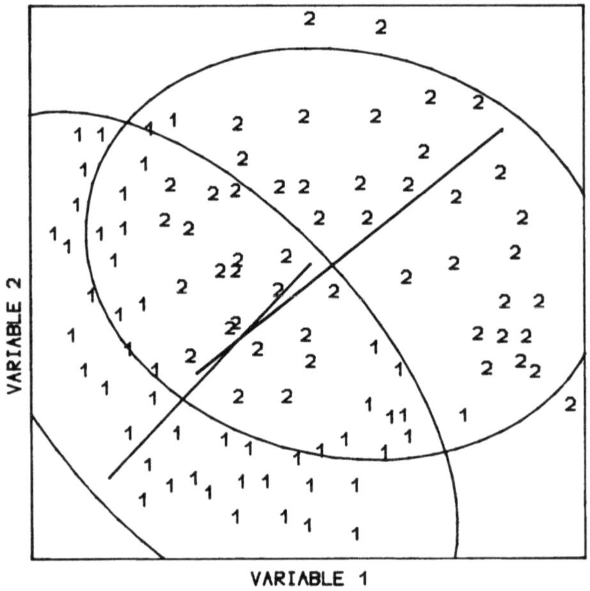

Fig. 28. Bivariate example of bayesian confidence ellipses and SIMCA one-component models in a case of complex distributions. The direction of SIMCA components is not the same as the main axes of the ellipses because of the separate scaling used in SIMCA

Although in this kind of contorted distribution it is possible to specify some subcategories by eigenvector projection and then to use BA or SIMCA with subcategory spaces, it seems desirable to have methods suitable for handling unsplit categories.

The evolution of KNN has produced methods based on potential functions, where each object is considered as an electric charge and the potential at a point is the sum of each individual contribution.

In the method of the software package ALLOC of Hermans and Habbema [53], the individual potential function has the form of a multidimensional normal probability function, and the overall function gives the conditional probability density of each category, so that the basic concepts of BA can be applied.

Coomans et al. [54, 55] have applied ALLOC to the differentiation of pure milk from different species and mixtures. A nonsymmetrical loss matrix was used in two-category classifications:

$$L = \begin{vmatrix} 0 & a \\ b & 0 \end{vmatrix},$$

where a is the loss associated with the misclassification of an object of class 2 into class 1, and b is the loss for the inverse misclassification. The ratio $b/(a + b)$ was used to build a family of discriminant boundaries between the categories. The above ratio has the significance of a threshold value of posterior probability for class 1, whereas $a/(a + b)$ is the threshold value for class 2. By a suitable choice of a and b a boundary zone is obtained and the objects which are in this zone are not classified into any class. So, ALLOC behaves almost as a class-modelling technique.

4.5 CLASSY

The method CLASSY attempts to bring together the appealing ideas of SIMCA and the Kernel density estimation of ALLOC (CLassification by ALLOC and SIMCA SYnergy) [44]. It has been applied to the classification of French wines [56] (Bourgogne and Bordeaux) by classical chemical and physical variables and by peak height of head-space chromatography.

In SIMCA the distribution of the object in the inner model space is not considered, so the probability density in the inner space is constant and the overall PD appears as shown in Figs. 29, 30 for the enlarged and reduced SIMCA models. In CLASSY, Kernel estimation is used to compute the PD in the inner model space, whereas the errors in the outer space are considered, as in SIMCA, uncorrelated and with normal multivariate distribution, so that the overall distribution, in the inner and outer space of a one-dimensional model, looks like that reported in Fig. 31. Figures 32, 33 show the PD of the bivariate normal distribution and Kernel distribution (ALLOC) for the same data matrix as used for Fig. 31. Although in the data set of French wines no really important differences have been detected between SIMCA (enlarged model), ALLOC and CLASSY, it seems that CLASSY should be chosen when the number of objects is large and the distribution on the components of the inner model space is very different from a rectangular distribution.

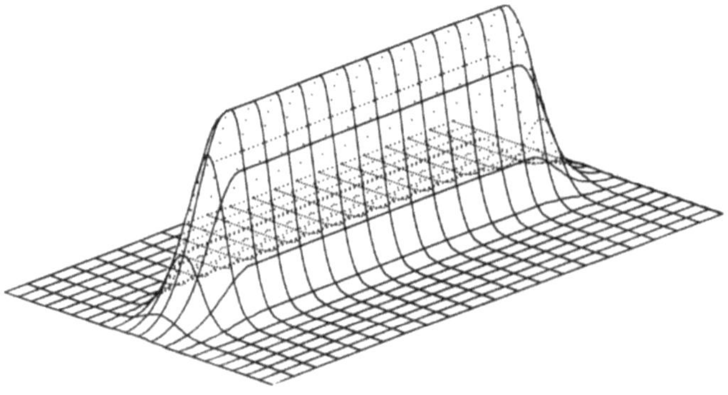

Fig. 29. Probability density function for SIMCA (usual enlarged model)

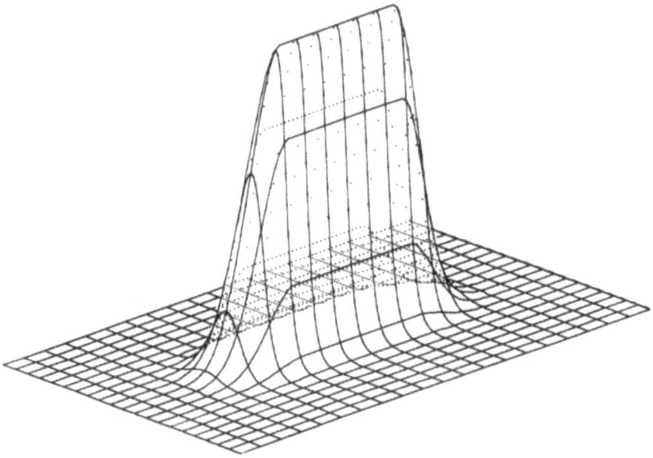

Fig. 30. Probability density function for SIMCA (reduced model)

On the other hand, when the number of objects is low (remember that SIMCA has been developed for this special case) the use of a Kernel estimation of probability density can have no significance, as shown by the example of Fig. 34, where the true distribution is a rectangular one.

4.6 Model Centred Bayesian Analysis

As SIMCA and ALLOC can be called the parents of CLASSY, BA and SIMCA are the parents of model centred bayesian analysis.

In SIMCA and CLASSY, the inner model space is formed from all the significant components. In practice, because of the difficulty of obtaining the number of significant components, often the dimensionality of the model is systematically examined to obtain the best number of components on the basis of method performances (prediction).

Model centred bayesian analysis (MCBA)[7] builds the inner model space only from the components that can be interpreted as due mainly to nonrandom underlying factors, determined by experimental design, or that show an almost rectangular distribution. In a study on Portuguese olive oils collected in the years 1975–1980[57], it was seen that the distribution on two eigenvectors (studied by the two-dimensional

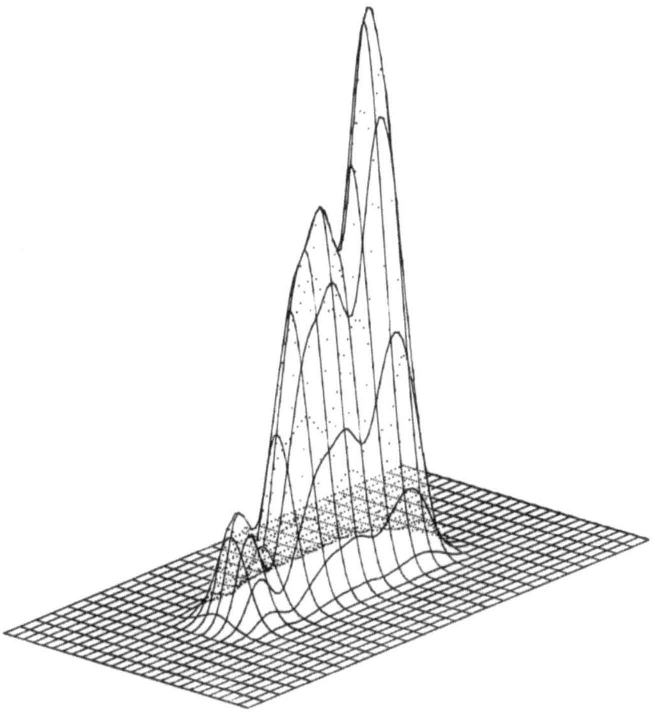

Fig. 31. Probability density function for CLASSY

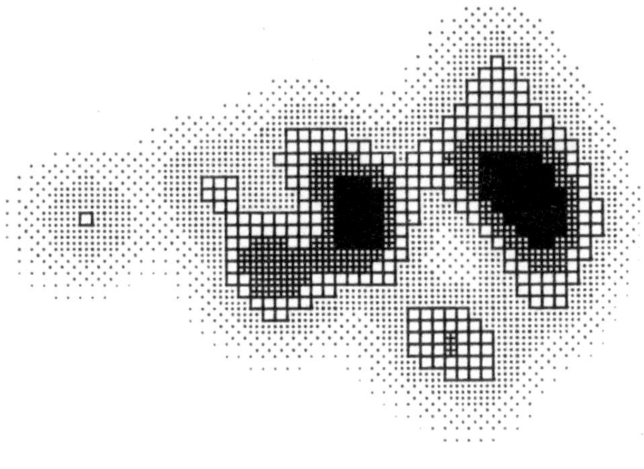

Fig. 32. Probability density function (contours) for ALLOC

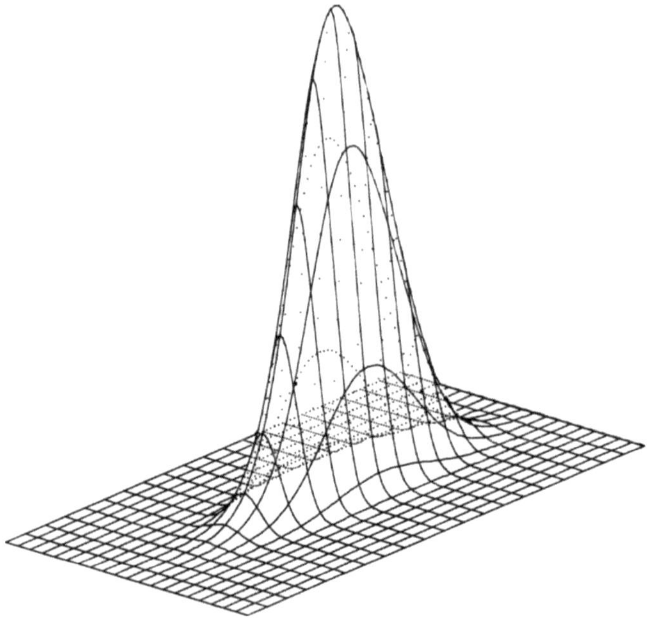

Fig. 33. Probability density function for Gaussian bivariate distribution

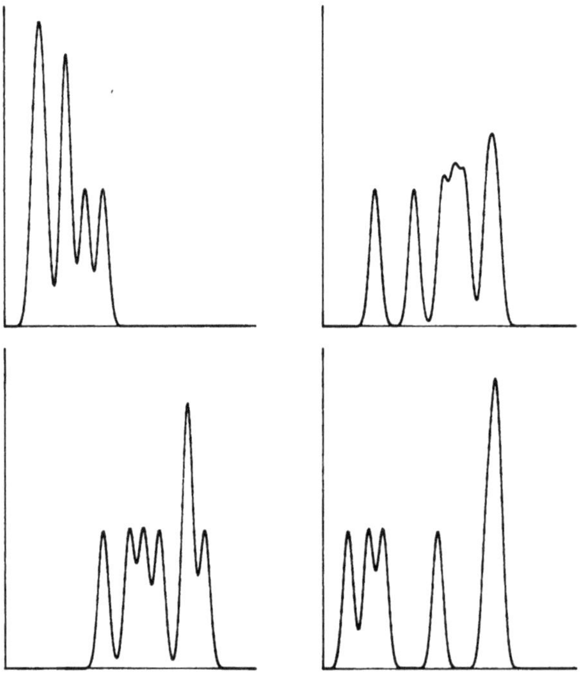

Fig. 34. Probability density function of CLASSY in the direction of the significant component for 4 random samplings of a few (8) objects from the same rectangular distribution

histograms of the scores) was very similar for each year. The position of objects in the plane of the two components appeared to be due to two climatic factors, the first connected with latitude, the second with the distance from the sea.

The sampling design (independent of the year) had been made on the basis of the "oil-producing districts", so that the distribution on the two components was determined by the geographical distribution of the districts and then by the experimental design. A random sampling, with constant sampling density, should produce, of course, a very different distribution.

The other components were independent of sampling design, and they collect the random errors with a multivariate normal distribution, where some significant correlations among variables are still present.

So, MCBA builds a covariance matrix of the residuals around the inner model and from this matrix it obtains a probability density function as bayesian analysis does, taking into account that the dimensionality of the inner space correspondingly reduces the rank of the covariance matrix from which a minor must be extracted.

Figure 35 shows the differences between the class spaces of BA (normal distribution), SIMCA and MCBA.

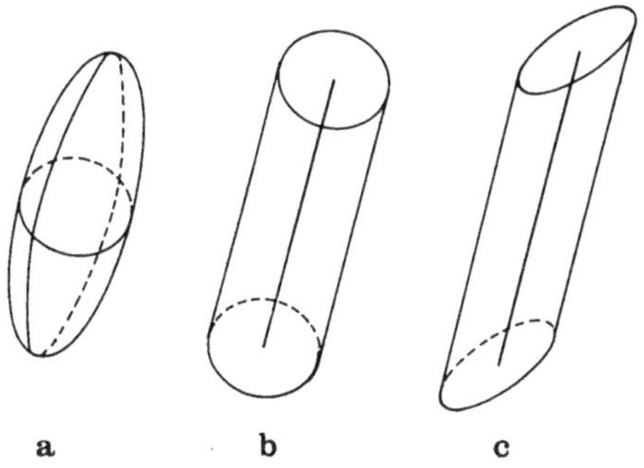

Fig. 35 a–c. Class spaces for bayesian analysis (**a**), SIMCA (**b**) and model centred bayesian analysis (**c**)

4.7 PRIMA

Pattern Recognition by Independent Multicategory Analysis (PRIMA) [58] is another method retaining some characteristics of SIMCA. It has been applied to the classification of brandies [59].

In PRIMA there is a separate scaling for each class: this kind of scaling, where each class has its scaling parameters, i.e., the mean and the standard deviation of each variable within the class, is commonly used in SIMCA, but in PRIMA it becomes a fundamental characteristic of the method. Then, under the assumption that the variables are uncorrelated, the PRIMA distance is the squared distance of an object from the class centroid. It has all the other characteristics of SIMCA.

So, PRIMA appears to be an oversimplified variant of SIMCA, with zero-dimensional inner space. The method can be applied when the class covariance matrices show very little correlation (no significant components can be detected).

In Table 1 the main differences between the parametric classification-modelling techniques are summarized. The large number of techniques suggested in the last few years and their use in the classification of food samples display the interest in

Table 1. Comparison of the parametric classification-modelling techniques

Method	Inner space	Outer space	Affected by scaling[a]	Provision for modelling and outlier detection
LDA	no	Multivariate normal correlated errors (pooled covariance matrix)	no	no
BA	no	Multivariate normal correlated errors (separate class covariance matrices)	no	yes
SIMCA	yes, constant probability density	Uncorrelated residuals	yes	yes
ALLOC	no	Kernel distribution	no	possible
CLASSY	yes, Kernel probability density	Uncorrelated residuals	yes	yes
MCBA	yes, constant probability density	Multivariate normal correlated residuals	yes	yes
PRIMA	no	Uncorrelated multivariate normal errors	no[b]	yes

[a] All methods based on the PC model are scaling dependent.
[b] Separate class scaling is a fundamental characteristic of this method.

identity food problems and that the ideal method cannot exist. Each problem requires a different method, according to its experimental design, its aims and the availability of computing facilities.

5 Clustering

Cluster analysis is the collective name for methods designed to understand the structure of a large data matrix, to recognize similarities among objects or among variables and to single out some categories as a set of similar objects (or variables).

Display methods (EP, NLM) can be considered as clustering techniques, when no apriori information is given about the subdivision of the dataset into categories. However, with the name of cluster analysis, we will denote the techniques working with the whole multivariate information in the following way.

Initially cluster analysis defines a measure of similarity, given by a distance or a correlation or the information content [60]. Distance can be measured as euclidean distance or Mahalanobis distance or Minkowski distance. Objects separated by a short distance are recognized as very similar, while objects separated by a great distance are dissimilar. The overall result of cluster analysis is reported as a dendrogram of the similarities obtained by many procedures.

Sometimes, in the field of food chemistry, display methods have been used to detect clusters, while clustering techniques have been used to confirm the subdivision into categories, and then as classification methods.

By EP, Forina and Tiscornia [61] were able to detect two categories within the Ligurian olive oils (East and West Liguria oils) and also two categories within the Sardinian olive oils (coast and inner Sardinia oils). In both cases, the detected categories were interpreted as connected to well-recognized climatic differences.

Lisle et al. [62] used the plots of original variables or ratios to detect clusters of rums, but the significance of these subcategories has not been explained. Likewise, Derde et al. [63] showed that West Liguria olive oils can be separated into two subcategories. However, it was not possible to give any explanation of these two subcategories because the original data matrix did not contain information about climate, soil or olive variety.

In practice, almost all studies on food have some prerecognized categories, and the detection of new categories in an eigenvector plot shows that some factors are unknown or that their importance has been underestimated, so that the classification problem has to be formulated again.

So, clustering techniques have been used for classification. Piepponen et al. [47] applied a hierarchical cluster analysis (CLUSTAN) to the classification of food oils (groundnut, soya, sunflower and maize) by their fatty acid composition. The dendrogram of the distances shows four well-separated clusters. Some suspect commercial samples of sunflower oil fall near the cluster of soya oils, so far from the claimed class that they cannot be considered genuine.

Aishima [64] used hierarchical cluster analysis on gas chromatographic profiles [10 peaks out of 93 measured peaks, 48 samples selected out of 200 samples of eight brands (categories) of soy sauce]. The obtained dendrograms were mainly discussed in connection with the results of linear discriminant analysis and the ten peaks selected for clustering.

Because of the long computing time of distances and the difficulty of evaluating the output dendrogram, cluster analysis is ususally performed with a small number of objects, as in the work of Aishima. However, Ooghe et al. [65] used cluster analysis with 269 objects in the study of French red wines by their amino acid spectrum.

The number of research applications of cluster analysis has shown a spectacular increase in recent years: the increased number of books and computer programs on this scientific tool will make it attractive to food chemists, so that they will be able to use cluster methods in true clustering problems.

6 Feature Selection

Feature selection, the choice of variables or features (combinations and transformations of original variables) relevant to the problem, is one of the most important aims of chemometrics. Reasons for feature selection are:

a) To simplify. When the number of variables or features is very great, many methods of pattern recognition cannot be applied because of memory storage requirements or very long computing times.

b) To improve classification ability, because variables that do not give useful discriminant information add noise, which is also shown by display methods, so darkening the useful information.
c) To save money and time, with the development of analytical control procedures which require the measurement of a subset of the originally selected variables.

6.1 Univariate Feature Selection

When feature selection is used to simplify, because of the large number of variables, methods must be simple. The univariate criterion of interclass variance/intraclass variance ratio (in the different variants called Fisher weights [37)], variance weights [37)] or Coomans weights [66)] is simple, but can lead to the elimination of variables with some discriminant power, either separately or, more important, in connection with other variables (Fig. 36).

The evaluation by these criteria of ratios and sums of variables can single out some relevant pairs of variables with an increased computing time. Van der Greef et al. [15)] selected 50 variables out of 420 original ones (mass spectra of urine

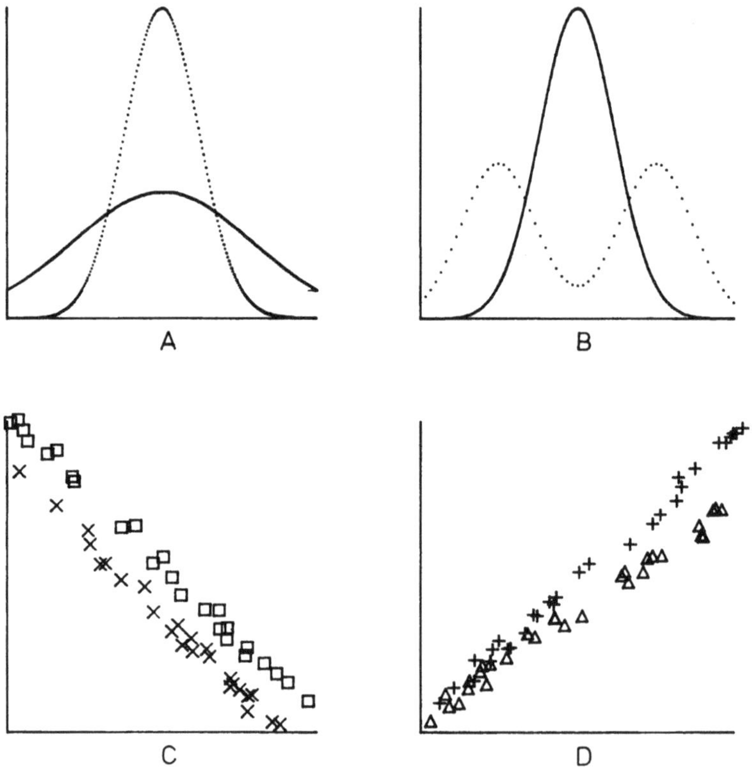

Fig. 36a–d. Some examples of single variables (**a, b**) or variable pairs (**c, d**) whose discriminant power cannot be detected by univariate Fisher weights. The transforms y + x (**a**) and y/x (**b**) have high univariate weight

and wine samples) by Fisher and Coomans weights: obviously the 263970 sums and ratios have not been considered.

Thus, the use of univariate criteria is advisable only when the number of variables is very large, or in a preliminary data analysis, because sometimes it is possible to find one or two variables that give enough information to solve the classification problem. In this way, Van der Greef et al. showed that Rhone and Bordeaux wines are almost completely separated in the plane of the masses 300 and 240.

Saxberg et al. [9] used Fisher weights in a classification problem where the whisky categories were standard Chivas (9 samples taken from a single used bottle opened many times in the air), commercial Chivas (new bottles from different batches) and non-Chivas. Six variables (chromatographic peak areas) were selected with a high Fisher weight for separating both standard Chivas from non-Chivas, and commercial Chivas from non-Chivas, but with a low weight for separating the two Chivas categories.

Eigenvector projection based on the six best variables is shown in Fig. 37, and it can be compared with the projection of Fig. 3 (obtained with 17 variables).

The separation between all Chivas and non-Chivas samples is not worsened. More, with the six retained variables the separation occurs along a single axis, because of the high correlation between the selected variables: really, also in this case only two variables are needed to obtain a perfect classification, and a line in the plane of these two variables (isoamyl alcohol and acetaldehyde) is an excellent discriminating function.

So, the important result of feature selection is a noticeable simplification of the analytical procedure required to detect counterfeit whisky.

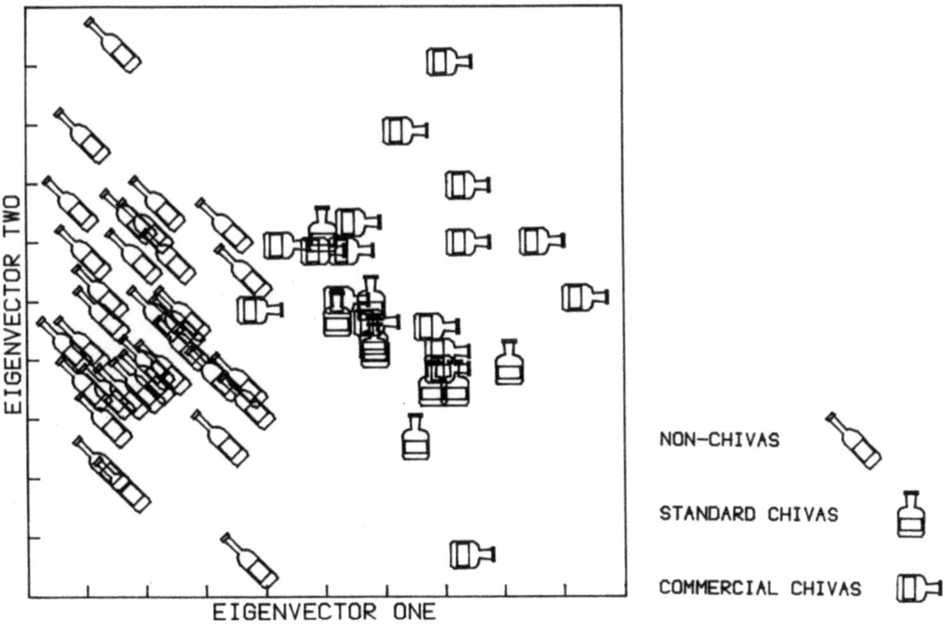

Fig. 37. Eigenvector plot of Chivas data after feature selection (6 selected variables). (Adapted from Ref. [9])

6.2 Feature Selection by Decorrelation (SELECT)

Because of the correlation between variables, univariate methods can select some variables that give the same information. The decorrelation method, used in the program SELECT of the software package ARTHUR, selects the first variable according to Fisher ratio, then this variable (let it be x_j) is subtracted from the remaining variables:

$$x_{iv} \rightarrow x_{iv} - \varrho x_{ij}$$

(where ϱ is the correlation between the generic variable v and the first selected variable j). Fisher ratios and correlation coefficients are recalculated for the non-selected variables, and a new best variable is obtained, which in turn is subtracted from the remaining variables, and so on, until the best Fisher ratio of the nonselected variables is less than a predetermined value.

Van der Voet and Doornbos [8] used this method in connection with the classification of French wines, and the result (prediction ability with the leave-one-out method) was as good as that with more sophisticated selection methods, but with a higher number of retained variables.

This method cannot solve distributions such as those of Fig. 36; however, because of its simplicity, we think that it can be recommended in preliminary data analysis, at least as an improvement in comparison with the univariate method.

6.3 Stepwise Selection by Linear Discriminant Analysis

In the same way as linear discriminant analysis is the most-used classification method, stepwise selection by LDA (SLDA) [28,29] is the selection method that shows the greatest number of applications in food chemistry.

Instead of the univariate Fisher ratio, SLDA considers the ratio between the generalized within-category dispersion (the determinant of the pooled within-category covariance matrix) and total dispersion (the determinant of the generalized covariance matrix). This ratio is called Wilks' lambda, and the smaller it is, the better the separation between categories. The selected variable is that that produces the maximum decrease of Wilks' lambda, tested by a suitable F statistic for the input of a new variable or for the deletion of a previously selected one.

The improvement of the separation obtained by this method has an average significance, i.e., it can happen that a better separation is due to two far categories that are further separated by a new selected variable; whereas two close categories are brought closer together, so that the overall classification rate can get worse.

For this reason, some criteria have been developed to try to select variables improving the separation of close categories (MAHAL, MAXMINF and MINRESID methods in the package SPSS). Scarponi and co-workers used all these selection methods in their work on the classification of Venetian white wines [33-35].

6.4 Feature Selection by SIMCA

In SIMCA, feature selection is carried out by the deletion of variables with both low modelling power and low discriminating power for each category. Modelling power measures the contribution of a variable to the model of a category, and discriminating power measures the contribution of a variable to the separation of categories.

This approach to feature selection has not been widely used in the classification problems of food chemistry, where sometimes SIMCA has been used after feature selection by decorrelation, SLDA or other methods [8].

6.5 Feature Selection by Prediction Ability

The ALLOC method with Kernel probability functions has a feature selection procedure based on prediction rates. This selection method has been used for milk [54,55] and wine [8] data, and it has been compared with feature selection by SELECT and SLDA. Coomans et al. [55] suggested the use of the loss matrix for a better evaluation of the relative importance of prediction errors.

A similar method, stepwise bayesian analysis, selects the variables giving the minimum number of classification plus prediction errors. When the error rate does not show further decrease, the procedure stops. The whole process is repeated with random subdivisions between the training and prediction sets. Only the variables that are selected independently of the subdivision are retained. This method has been used [19] with the data set of Fig. 4; only 5 variables were selected and a very high prediction ability was obtained.

7 Correlation and Optimization

The methods of quantitative chemometrics have been widely applied to the field of multivariate calibration, where some excellent reviews and books have been published recently [1,2,67]. We have found some applications to the study of the relationships between chemical and sensory variables (Fig. 38) [68]. All the other possibilities (itemized in Sect. 1) have not yet been explored.

The diffusion of correlation methods and related software packages, such as partial-least-squares regression (PLS), canonical correlation on principal components, target factor analysis and non-linear PLS, will open up new horizons to food research.

All these methods search for significant correlations between two blocks, X and Y, of variables, so that the obtained relationship has predictive value and gives a possibility of interpretation.

At first, optimization methods are not interested in the relationships between the two blocks of variables, but only in the X value that produces the best value of Y (often this Y block contains only one variable), according to some kind of requirement. This is generally a search for a maximum (or a minimum) of the hypersurface $y = f(X)$. Obviously, when the equation of this surface is known, this maximum search is much easier, so that the spread of correlation techniques will be profitable to otpimization problems too.

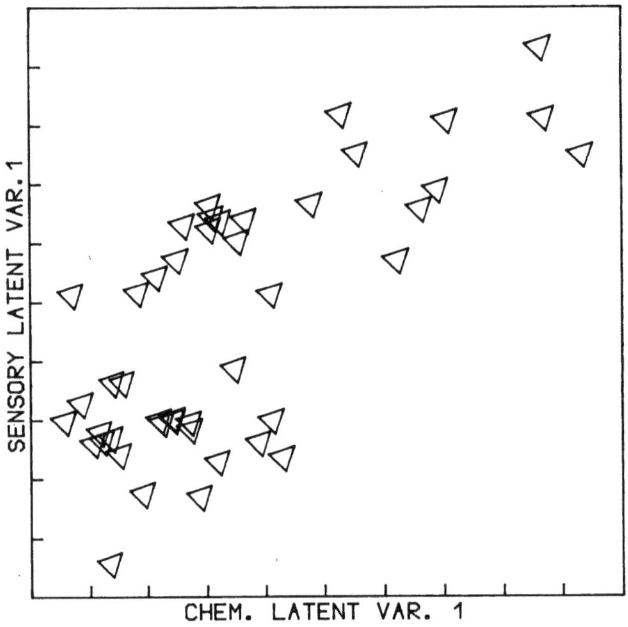

Fig. 38. Partial least-squares correlation between chemical and sensory variables (40 chemical and 4 sensory variables). The selected chemical variables explain 45% of the variance of the sensory block. (Adapted from Ref. [68])

Moll et al. [69] searched for the optimum blending of malts when they are mashed. Because of the possibility of nonlinear interactions, a chemical quantity in the mixture cannot be simply considered a linear combination of each constituent. In the case of three malts, a full factorial design is used to obtain, by seven determinations, the seven coefficients a of the equation

$$y = a_1x_A + a_2x_B + a_3x_C + \dots + a_7x_Ax_Bx_C ,$$

where x_A, x_B and x_C are the percentages of the three malts. In this case the above equation was checked and used to find the combination giving the required value of the chemical quantity Y.

When the number of factors is too high, or when the equation is too complex to be obtained by few determinations, fractional factorial design [70] and simplex movement on the response surface toward the optimum set of conditions [71] are the methods of choice, to avoid too large a number of experimental determinations and the effect of insignificant variations.

8 Conclusion

This review of chemometrics in food chemistry does not consider all the papers on this subject published during recent years, mainly because many papers appeared in

Table 2. Applications of multivariate methods in the classification of wines

Categories	N. of samples	Variables	N. of selected features	Methods	Classification ability [%]	Ref.
Asti Freisa	56	Tartaric		LDA		72)
Asti Barbera	56	acid,				
		N total,				
		Methanol				
Total	112	3				
Moselle 1970	10	Total acids,	3	EP		73)
Moselle 1971	16	5 alcohols,	3	SIMCA	79–80	
Rhine 1970	14	7 elements,		KNN	60–75	
Rhine 1971	9	etc.				
Total	49	16				
Riesling Muscat		27 aroma		SLDA		74)
Morio Muscat		compounds				
Total		27				
Riesling	14	7 elements		EP		75)
White Pinot	13					
Red Oltrepó,						
Pavese	15					
Bonarda	21					
Total	63	7				
Pinot noir from:						76)
France	14	17 elements	3	EP		
California	9			SIMCA		
Pacific Northwest	17			KNN		
Total	40	17				
Pinot noir from:						77)
France	14	137 organic	2	SIMCA	81–98	
California	9	compounds		KNN	77–90	
Pacific Northwest	17					
Total	40	137				
Pinot noir from:						78)
France	14	17 elements,		PCA		
California	9	137 organic				
		substances,				
Pacific Northwest	17	14 sensory				
		scores				
Total	40	168				
White Riesling	11	27 volatile	19	PCA		79)
Chardonnay	9	components		SLDA		
French Colombard	4					
Total	24	27				
Pinotage	10	10 esters,	2	SLDA		21)
Cabernet		4 higher				
Sauvignon	10	alcohols,	2			
		2 acids	2			
Total	20	16				

Table 2. (continued)

Categories	N. of samples	Variables	N. of selected features	Methods	Classification ability [%]	Ref.
Bourgogne	101	26 amino		Clustering		65)
Bordeaux	72	acids				
Rhône	12					
Languedoc	21					
Other wines	63					
Total	269	26				
Soave	14	4 elements,	6	SLDA	83–100	80)
Prosecco	14	etc.				
Verduzzo	14					
Total	42	10				
Prosecco 79	19	4 elements,	5	SLDA	82–93	33)
Prosecco 77	14	etc.				
Total	33	10				
Bordeaux	11	Mass spectrum	50	EP		15)
Rhône	11	of volatile		NLM		
		compounds		KNN		
Total	22	420				
Soave	18	5 elements,	9	SLDA	89–97	81)
Prosecco	33	etc.	6	KNN	88–96	
Verduzzo	20					
Total	71	19				
Muller-Thurgau	20	Volatile	40	PCA		82)
Riesling	25	compounds,		NLM		
		amino acids		KNN		
				SIMCA		
Total	45	115				
Bordeaux	21	3 elements,	6	SLDA	88–95	8)
Bourgogne	19	4 organic	11	LLM	78–98	
		acids, pH,	4	KNN	83–95	
		etc.	8	ALLOC	80–98	
				SIMCA	80–93	
Total	40	20				
Soave	18	5 elements,	11	SLDA	95–100	83)
Prosecco	33	pH, etc.		KNN	82–85	
Verduzzo	20					
Tocai	21					
Total	92	19				
Verduzzo	18	7 ethyl-	11	SLDA	90–100	34)
Soave	10	esters,		KNN	70–73	
Tocai	11	5 alcohols,				
		etc.				
Total	39	22				
Soave	32	4 elements,		SLDA	71–97	35)
Prosecco	47	etc.		KNN	74–94	
Verduzzo	35					
Total	114	10				

Table 2. (continued)

Categories	N. of samples	Variables	N. of selected features	Methods	Classification ability [%]	Ref.
Barbera from:		10 elements,		EP		[38]
Oltrepó Pavese	6	pH, etc.		LDA		
Piemonte	6					
Veneto	1					
Total	13	20				
Bordeaux	21	11 peaks		SIMCA		[56]
Bourgogne	19	from head-		CLASSY		
		-space chrom.,		ALLOC		
		etc.				
Total	40	31				
Pinot noir from:		17 elements,		PLS		[84]
Pacific Northwest	17	137 organic				
California	9	substances,				
France	14	14 sensory				
		scores				
Total	40	154				
White wines	47	4 sensory		PLS		[68]
		scores,				
		40 chemical				
		substances				
Total	47	44				
White wines	47	4 sensory		FA		[85]
		scores,				
		40 chemical				
		substances				
Total	47	44				
Verduzzo	35	10 head-		SIMCA		[86]
Soave	21	-space		KNN	71–91	
Tocai	21	components				
Total	77	10				
Rosés	15	13 phenolic		LDA	93–100	[87]
Reds	18	compounds				
Clarets	15					
Total	48	13				
Barolo	59	Alcohol,	8	EP		[50]
Barbera	71	pH, K, Ca,		LDA	97–98	
Grignolino	48	flavanoids,		SIMCA	96–100	
		proline, etc.		BA	96–100	
Total	178	28				

journals that the authors were not able to find. Moreover, the attention given to the various chemometrical topics has been made roughly proportional to the number of applications and to the innovations made in connection with food chemical problems. So, the possible applications are numberless, much more than the examples we have just described. Really, all the main human activities are multivariate classifications and correlation, because each human sense receives a great deal of simultaneous information from each object observed, touched, smelled or tasted. Since each "food object" studied by chemistry is destined for a final "multivariate" user, the univariate approach can be justified only by the limits imposed by the available instrumentation and funds. Multivariate strategies, however, suffer from limits of the same kind: nobody has determined all the chemical contituents in a food sample.

In Table 2 many of the applications of multivariate methods in the classification of wines are reported: a great many different classes of variables have been used. The kind of variable ranges from the results of classical chemical determinations to trace elements, esters, aminoacids, head-space chromatographic peaks and mass spectra peaks. The number of variables ranges from 3 to 400 and the number of objects from about 10 to some hundreds. Some sets of variables have been recognized as very useful in classification and very interesting results heve been obtained on the correlation with sensory scores. However, these findings are only a partial answer to the general problem, because each research team has considered only a small number of the chemical species in the wine samples.

For long-term application of the obtained results, much more additional work is required. An exhaustive experimental design, a correct sampling, the selection and the method of measurement of the variables, the study of the correlations with sensory scores and consumer references, and the choice of chemometrical methods require interdisciplinary efforts of oenologists, chemists and chemometricians. A project proposal for the authentication of wines has been prepared, with ten classes of variables (from classical analytical data to trace elements and isotopic abundance) to be measured in 20 laboratories (with interlaboratory investigation) on some hundred samples, so connecting all the necessary experience and instruments. These laboratories are in many wine-producing and wine-importing countries of the European Economic Community. The authentication of foods is valid only when controlled by the international community of food scientists.

This project is very ambitious, because, besides the practical results, it aims at defining a methodology. It is expensive because of the great amount of analytical and data-processing work, but it will define the few relevant chemical species to be determined in real applications. It is futurist, because at present only exceptional food control laboratories can routinely measure all the classes of chemical variables. But, we have seen the astounding development of analytical instruments and methods during the last twenty years. In the near future, these instruments will show increased availability, and new techniques and methods will appear. Food chemistry is working to prepare this future, where physical instruments will give more and more data and chemometrical methods will extract the relevant chemical information.

9 References

1. Martens, H., Naes, T.: Trends Anal. Chem. *3*, 204 (1984)
2. Naes, T., Martens, H.: ibid. *3*, 266 (1984)
3. Brown, P. J.: J. R. Stat. Soc. B. *44*, 287 (1982)
4. Grotch, S. L.: Three-Dimensional Graphics for Scientific Data Display and Data Analysis, in: Chemometrics, Mathematics and Statistics in Chemistry (Kowalski, B. R., ed.), p. 439, Reidel, Dordrecht 1984

5. Malinowski, E. R., Howery, D. G.: Factor Analysis in Chemistry, Wiley Interscience, New York 1980
6. Wold, S.: Technometrics *20*, 397 (1978)
7. Forina, M., Armanino, C.: Classification and Modelling Techniques in Food Chemistry, in: COMPANA 85, (Danzer, K., ed.), p. 89, University Jena Series, Jena 1986
8. Van der Voet, H., Doornbos, D. A.: Anal. Chim. Acta *159*, 159 (1984)
9. Saxberg, B. E. H., Duewer, D. L., Booker, J. L., Kowalski, B. R.: ibid. *103*, 201 (1978)
10. Forina, M., Lanteri, S.: Data Analysis in Food Chemistry, in: Chemometrics, Mathematics and Statistics in Chemistry (Kowalski, B. R., ed.), p. 305, Reidel, Dordrecht 1984
11. Benzécri, J. P.: L'Analyse des Données, Editions Dunod Paris, 1984
12. Lewi, P. J.: Multivariate Data Analysis in Industrial Practice, Chemometrics Research Studies Press, Wiley, Chichester 1982
13. Thielemans, A., Massart, D. L.: The Effect of Different Scaling Techniques and/or Transformations on Display Results: Application to Food Data, Anal. Chim. Acta, in press
14. Kowalski, B. R., Bender, C. F.: J. Am. Chem. Soc. *95*, 686 (1973)
15. Van der Greef, J., Tas, A. C., Bouwman, J., Ten Noever de Brawuw, M. C., Schreurs, H. P.: Anal. Chim. Acta *150*, 45 (1983)
16. Derde, M. P., Massart, D. L., Ooghe, W., De Waele, A.: J. Autom. Chem. *3*, 136 (1983)
17. Forina, M., Armanino, C., Lanteri, S., Calcagno, C.: Ann. Chim. (Rome) *73*, 641 (1983)
18. Forina, M., Armanino, C.: ibid *72*, 127 (1982)
19. Forina, M., Lanteri, S., Castino, M., Leardi, R.: in: Atti VI Congresso Nazionale (Zambonin, P. G., ed.), p. 70, SCI, Bari 1985
20. Gilbert, J., Shepherd, M. J., Wallwork, M. A., Harris, R. G.: J. Apic. Res. *20*, 125 (1981)
21. Marais, J., Van Rooyen, P. C., Du Plessis, C. S.: S. Afr. J. Enol. Vitic. *2*, 19 (1981)
22. Reiner, L., Piendl, A.: Brauwissenschaft *27*, 2 (1974)
23. Gaydou, E. M., Ralambofetra, E., Rakotovao, L., Llinas, J. R.: Analusis *13*, 379 (1985)
24. Smeyers-Verbeke, J., Massart, D. L., Coomans, D.: J. Assoc. Off. Anal. Chem. *60*, 1382 (1977)
25. Aishima, T.: J. Food Sci. *47*, 1562 (1982)
26. Forina, M., Armanino, C., Lanteri, S.: Riv. Soc. Ital. Sci. Alim. *11*, 15 (1982)
27. Kawahara, F. K., Santner, J. F., Julian, E. C.: Anal. Chem. *46*, 266 (1974)
28. Nie, N. H., Hull, C. H., Jenkins, J. C., Steinbrenner, K., Bent, D. H.: Discriminant Analysis, in: Statistical Package for the Social Sciences, p. 447, McGraw-Hill, New York 1975
29. Jenrich, R. I.: Stepwise Discriminant Analysis, in: Statistical Methods for Digital Computers (Enslein, K., Ralston, A., Wilf, H. S., eds.) p. 76, Wiley, New York 1960
30. Chernoff, H.: J. Am. Stat. Assoc. *68*, 361 (1973)
31. Yeung, E. S.: Anal. Chem. *52*, 1120 (1980)
32. Andrews, H. C.: Mathematical Techniques in Pattern Recognition, Wiley, New York, 1972
33. Scarponi, G., Moret, I., Capodaglio, G., Cescon, P.: J. Agric. Food Chem. *30*, 1135 (1982)
34. Moret, I., Scarponi, G., Cescon, P.: J. Sci. Food Agric. *35*, 1004 (1984)
35. Moret, I., Di Leo, F., Giromini, V., Scarponi, G.: J. Agric. Food Chem. *32*, 329 (1984)
36. Wencker, D., Louis, M., Patris, A., Laugel, P., Hasselmann, M.: Analusis *9*, 498 (1981)
37. Harper, A. M., Duewer, D. L., Kowalski, B. R., Fasching, J. L.: ARTHUR and Experimental Data Analysis: the Heuristic Use of a Polyalgorithm, in: Chemometrics Theory and Application (Kowalski, B. R., ed.), p. 14, ACS Symposium Series 52, Washington, D. C. 1977
38. Baldi, M., Riganti, V., Specchiarello, M.: Industrie delle Bevande *6*, 1 (1984)
39. Derde, M. P., Massart, D. L.: UNEQ: A Disjoint Modelling Technique for Pattern Recognition based on the Normal Distribution, Anal. Chim. Acta, in press
40. Ames, A. E., Szonyi, G.: How to Avoid Lying with Statistics, in: Chemometrics Theory and Application (Kowalski, B. R., ed.), p. 219, ACS Symposium Series 52, Washington, D. C. 1977
41. Wold, S.: in: Textes des Communications of First International Symposium on Data Analysis and Informatics, p. 683, IRIA, Le Chesnay 1977
42. Wold, S., Albano, C., Dunn III, W. J., Esbensen, K., Hellberg, S., Johansson, E., Sjostrom, M.: Pattern Recognition: Finding and Using Regularities in Multi-Variate Data, in: Food Research and Data Analysis (Martens, H., Russwurm, H., eds.), p. 147, Applied Science Publ., Barking 1983

43. Wold, S., Albano, C., Dunn III, W. J., Edlund, U., Esbensen, K., Geladi, P., Hellberg, S., Johansson, E., Lindberg, W., Sjostrom, M.: Multivariate Data Analysis in Chemistry, in: Chemometrics, Mathematics and Statistics in Chemistry (Kowalski, B. R., ed.), p. 17, Reidel, Dordrecht 1984
44. Van der Voet, H., Doornbos, D. A.: Anal. Chim. Acta *161*, 115 (1984)
45. Stenroos, L., Siebert, K. J.: J. Am. Soc. Brew. Chem. *42* (2), 54 (1984)
46. Bisani, M. L., Clementi, S.: Chemometrics in Food Chemistry: Classification of Camomiles from Central Italy, in: Food Research and Data Analysis (Martens, H., Russwurm, H., eds.), p. 428, Applied Science Publ., Barking 1983
47. Piepponen, S., Vainionpaa, J., Roos, A.: Kem.-kemi *9*, 397 (1982)
48. Forina, M., Armanino, C., Lanteri, S., Tiscornia, E.: Classification of Olive Oils from their Fatty Acid Composition, in: Food Research and Data Analysis (Martens, H., Russwurm, H., eds.), p. 189, Applied Science Publ., Barking 1983
49. Derde, M. P., Coomans, D., Massart, D. L.: Application of Class Modelling Techniques for the Classification of Olive Oils, in: Food Research and Data Analysis (Martens, H., Russwurm, H., eds.), p. 427, Applied Science Publ., Barking 1983
50. Forina, M., Armanino, C., Castino, M., Ubigli, M.: Multivariate Data Analysis as a Discriminating Method of the Origin of Wines, Vitis *25*, 189 (1986)
51. Coomans, D.: Patroonherkenning in de Medische Diagnose Aan de Hand van Klinische Laboratorium Ondrzoeken, Pharm. Thesis, Vrije Universiteit Brussel 1982
52. Jurs, P. C., Kowalski, B. R., Isenhour, T. L.: Anal. Chem. *41*, 690 (1969)
53. Hermans, J., Habbema, J. D.: Manual for the ALLOC-Discriminant Analysis Program, Dept. Medical Statistics, Univ. Leiden 1976
54. Coomans, D., Derde, M. P., Massart, D. L., Broeckaert, I.: Anal. Chim. Acta *133*, 241 (1981)
55. Coomans, D., Massart, D. L., Broeckaert, I.: ibid. *134*, 139 (1982)
56. Van der Voet, H., Doornbos, D. A.: ibid. *161*, 125 (1984)
57. Forina, M., Armanino, C., Lanteri, S., Calcagno, C., Tiscornia, E.: Riv. Ital. Sostanze Grasse *9*, 607 (1983)
58. Juricskay, I., Veress, G. E.: Anal. Chim. Acta *171*, 61 (1985)
59. Veress, G. E.: private communication
60. Massart, D. L., Kaufman, L.: The Interpretation of Analytical Chemical Data by the Use of Cluster Analysis, Wiley, New York 1983
61. Forina, M., Tiscornia, E.: Ann. Chim. (Rome) *72*, 143 (1982)
62. Lisle, D. B., Richards, C. P., Wardleworth, D. F.: J. Inst. Brew., London *84*, 93 (1978)
63. Derde, M. P., Coomans, D., Massart, D. L.: Anal. Chim. Acta *141*, 187 (1982)
64. Aishima, T.: Instrum. Anal. Foods *1*, 37 (1983)
65. Ooghe, W., Kastelijn, H., De Waele, A.: Ann. Falsif. Expert. Chim. *74*, 381 (1981)
66. Varmuza, K.: Pattern Recognition in Chemistry, Lecture Notes in Chemistry, Vol. 21, Springer-Verlag, Berlin Heidelberg 1980
67. Martens, H.: Multivariate Calibration, Dr. techn. thesis, Technical University of Norway, Trondheim 1985
68. Bertuccioli, M., Clementi, S., Giulietti, G.: Vini Ital. *26*, 27 (1984)
69. Moll, M., Flayeux, R., Mathieu, D., Phan Tan Lou, R.: J. Inst. Brew., London *88*, 139 (1982)
70. Hunter, J. S.: Experimental Design: Fractionals, in: Chemometrics, Mathematics and Statistics in Chemistry (Kowalski, B. R., ed.), p. 1, Reidel, Dordrecht 1984
71. Deming, S. N.: Linear Models and Matrix Least Squares in Clinical Chemistry, in: Chemometrics, Mathematics and Statistics in Chemistry (Kowalski, B. R., ed.), p. 1, Reidel, Dordrecht 1984
72. Castino, M.: Atti Accad. Ital. Vite Vino *27*, 1 (1975)
73. Kwan, W. O., Kowalski, B. R.: J. Food Sci. *43*, 1320 (1978)
74. Rapp, A., Hastrich, H., Engel, L., Knipser, W.: Flavor of Foods and Beverages, Academic Press, London 1978
75. Silva, A.: Industrie delle Bevande *2*, 55 (1980)
76. Kwan, W. O., Kowalski, B. R., Skogerboe, R. K.: J. Agric. Food Chem. *27*, 1321 (1979)
77. Kwan, W. O., Kowalski, B. R.: ibid. *28*, 356 (1980)
78. Kwan, W. O., Kowalski, B. R.: Anal. Chim. Acta *122*, 215 (1980)

79. Noble, A. C., Flath, R. A., Forrey, R. R.: J. Agric. Food Chem. *28*, 346 (1980)
80. Scarponi, G., Moret, I., Capodaglio, G.: Riv. Vitic. Enol. *34*, 254 (1981)
81. Moret, I., Scarponi, G., Cescon, P.: Rass. Chim. *35*, 319 (1983)
82. Schaefer, J., Tas, A. C., Velisek, J., Maarse, H., ten Noever de Brauw, M. C., Slump, P.: in: Proceedings of the Third International Flavour Conference, p. 1, Corfu 1983, available at TNO postbus 360, 3700 AJ Zeist, The Netherlands
83. Moret, I., Scarponi, G., Capodaglio, G., Giromini, V., Cescon, P.: Ann. Chim. (Rome) *74*, 73 (1984)
84. Frank, I., Kowalski, B. R.: Anal. Chim. Acta *162*, 241 (1984)
85. Bertuccioli, M., Bracardi, M., Clementi, S.: Industrie delle Bevande *7*, 19 (1985)
86. Moret, I., Scarponi, G., Capodaglio, G., Cescon, P.: Riv. Vitic. Enol. *38*, 254 (1985)
87. Santa-Maria, G., Garrido, J. L., Diez, C.: Lebensm. Unters. Forsch. *182*, 112 (1986)

Species Identification for Trace Inorganic Elements in Biological Materials

Philip H. Ekow Gardiner

Institut für Angewandte Physikalische Chemie, Kernforschungsanlage Jülich, Postfach 1913, 5170 Jülich, FRG

Table of Contents

The techniques and methods that could be applied to chemical speciation in biological systems are surveyed and the limitations are highlighted. In addition, changes that occur in the samples that may have a detrimental effect on the results are examined.

1 Introduction

The extent and type of interaction of an element with other constituents in a system determine its chemical behaviour and provide the basis for our understanding of the various biochemical processes in which it participates. It is therefore apparent that a better appreciation of the biochemical role of the element can be gained when the various forms in which it occurs are identified. The identification of the various physicochemical forms of an element or chemical speciation, as it is now called, presents many challenges. Firstly, the levels of some species are so low that more stringent control of the extraneous sources of contamination and more sensitive methods of detection are required than is the case for the total determination. Secondly, in order to obtain results that accurately reflect the speciation of an element, realistic assumptions have to be made about the system and the appropriate experimental conditions must be chosen. This requirement can only be satisfied when enough is known about other chemical constituents that make up the system; unfortunately, this information is not always available. Thirdly, for the total characterisation of a species, a wide range of techniques are required, and these may not be availiable in a single laboratory. Moreover, an interdisciplinary approach may, in some cases, be needed in order to plan and execute the experiments. Finally, relevant information about the changes that occur in a sample either after collection and/or during storage should be available. However, this aspect is often ignored by most workers, and as a result the success or failure of some experiments depends on the intuition and experience of the investigator.

In this article some of the methods and techniques that have been used for the speciation of trace inorganic elements in biological materials, and the approaches used to solve some of the above problems will be discussed.

2 Sample Collection, Pretreatment and Storage

The major consideration in the choice of techniques and methods used for speciation studies is that the integrity of the species should be maintained. In other words, the interactions between the element under study and other constituents directly associated with it should not be disrupted. Since the strength of these interactions can range from weak van der Waals type force to strong covalent bonds, it is to be expected that the conditions under which the stability of each type of interaction is affected would vary. However, the species are more likely to remain intact when the experimental conditions closely resemble those found in vivo. Generally, changes in pH, temperature, ionic strength, ionic composition and partial pressure during sampling, storage or fractionation could adversely affect the distribution of the species. The effect of changes in the above parameters on the stability of the various types of interactions is discussed later.

The possibility of contaminating the sample with the elements under investigation during the various steps is a major risk in trace analysis. The precautions required in order to minimize this problem are similar to those already identified when dealing with biological materials used for total element determinations. Heydorn [1], Versieck et al. [2], Stoeppler [3], Behne [4], and Aitio et al. [5] have discussed the extent of the problem and the necessary precautions. However, some problems that are of direct relevance to speciation will be discussed below.

A complication in the study of biological materials is that the species of an element which are either formed and/or take part in physiological reactions in different compartments may vary. It is therefore necessary, especially during the sampling and pretreatment steps, that the constituents from different compartments are very carefully separated. For example, in the study of the speciation of an element in blood serum contaminants originating from the erythrocytes could lead to errors in the results.

Another important consideration is the effect of the reactions that continue, albeit at different rates, and in some cases along new pathways, after sampling. These reactions could introduce new products, modify existing species and cause the loss of volatile components. Consequently, the speciation profile of some elements may be changed. In order to minimize this effect, it is essential to process the samples as soon as possible after collection.

The addition of anticoagulents and/or preservatives, a practice that may be tolerated for total element determination, should be avoided for two reasons. First, the compounds used are usually complexing agents. They therefore could bind various trace elements. Second, they could destroy some species. For example, the addition of potassium dichromate to urine in the presence of nitric acid could destroy methylmercury [6]. If the use of such compounds is unavoidable, then there should be experimental evidence to show that the speciation of the element under study has not been adversely affected.

Besides the above factors that should be taken into consideration when dealing with biological materials used for speciation, the procedures used for sampling are similar to those applied when only the total element content of the samples is of interest. This statement is also true for the pretreatment steps. However, in order to preserve the interaction between the element under study and other constituents associated with it, the constraints placed on the choice of procedures are more severe. Furthermore, the techniques and methods used are chosen in the light of relevant information about the known chemical behaviour of the element in its various binding forms or associations. A working knowledge of the methods required for the isolation of the various biochemical compartments, that may be of interest to the investigator, is also necessary.

An understanding of the chemical behaviour of the element can aid in the choice of appropriate techniques and methods, the application of which would not disrupt the interaction of the element with associated constituents. For example, in the study of aluminium some relevant information may include its amphoteric nature, its ability to form predominantly ionic complexes, its tendency to form hydroxides, and the stability of aluminium complexes formed with biological ligands. It is clear that in order to maintain the ionic interactions the pH, ionic strength and, of lesser importance, the ionic composition of the medium used for sample preparation should be similar to that found in vivo. In addition, highly charged surfaces should not come into contact with the sample.

On the other hand, information on copper would show its ability to participate in predominantly covalent bonding, its tendency to take part in redox reactions and the ability to form coloured complexes, in which the copper to ligand ratio is a constant.

Here, small changes in ionic strength and pH could be tolerated. However, the addition of reducing or oxidising agents to the solution, and/or the buildup of redox couples should be avoided. This type of background information about an element is required not only to enable the choice of appropriate experimental conditions but also for the interpretation of the results.

The complexity of the pretreatment procedure chosen is largely determined by the nature of the sample and the information required. Most present-day analytical instruments are designed to process gas and/or liquid samples. These could, therefore, be used with the minimum of pretreatment. Solid samples, on the other hand, have to be brought first into solution. This can be accomplished by homogenisation of the sample. However, as already emphasised it is important to separate the various compartments.

For example, if the purpose of the experiment is to study the speciation of an element in a specific cell type in a given tissue it is essential that all the extracellular fluid is removed from the tissue before the homogenisation step. This step could be followed by ultracentrifugation of the homogenate in order to separate the cell under study from others that may be present in the medium. An estimate of the degree of purity of the end product should be given. The harvested cells may then be further processed.

It is not possible to prescribe specific pretreatment procedures here because these can only be decided upon when the system and the purpose of the experiments has been properly defined. However, a wealth of information exist in various biochemical reference books [7,8)] on how to isolate various biological compounds. The recommended techniques and methods could be used as part of the trace element speciation protocol often after slight modification, taking into consideration the following points: First, the trace element blank levels have to be low, less than 10% of the total concentration in the sample. Second, the reagents used should not interfere with subsequent analytical determinations. Third, the experimental conditions should not deviate markedly from those found in vivo, especially the pH and ionic strength of the medium.

The problems arising from the storage of biological materials, as pointed out earlier, have been largely ignored by most investigators. The information available in the literature deals with the effect of storage on the total element content (5). These studies have indicated that for some elements temperatures below about —20 °C are required in order to avoid losses. Other sources of loss are through precipitation, adsorption on the container surfaces and the evaporation of the sample constituents. Unfortunately, the addition of acid or other perservatives, which may reduce some effects, cannot be used for reasons already given above. It is possible that a factor that may determine the behaviour of an element during storage is its speciation in the sample.

During storage there are changes in the sample that may have some consequence for the speciation of an element. First, the complex three-dimensional structure of most biological molecules may be destroyed. This may lead to the loss of enzymatic activity, and in some cases the associated element may be lost. Second, the natural proteolysis and/or autolysis reactions could result in the breakdown of the molecules, and as a consequence the results of the speciation experiments may indicate that the element is associated with only a fragment of the original molecule. Indeed, it could also be possible that enzymes or microorganisms present in the sample may also produce new species. The extent of the damage to the molecules depends on the storage conditions. There are molecules that could be stored for reasonable lengths of time without discernable changes; however, some molecules are so labile that the samples have to be used immediately after collection.

The changes in the three-dimensional structures of the biological molecules with storage, and the length of time that a sample can be stored without adverse effects are points that have to be addressed in any future study on the effect of storage on the chemical speciation of elements.

3 The Stability of Metal-containing Species

Before discussing the factors that affect the stability of various metal-containing species it is essential to draw some comparision between non-living and living systems. This can best be done by a hypothetical example.

Let us consider an aqueous system at a given pH, temperature and ionic strength which contains various ligands but for the sake of simplicity only one metal ion. It is possible given the concentrations of the ligands and metal, and a knowledge of the stability constants of possible species to calculate the distribution of the various metal-containing species with a reasonable degree of accuracy. The level of accuracy will depend on the available relevant information about the system, the purity of the reagents and the effect of side-reactions. The amount of each species will depend on the concentration of each ligand and its relative chemical affinity for the metal. A measure of this affinity is given by the change in free energy of the reaction between a given ligand and the metal. It is therefore possible by using purely thermodynamic considerations to predict which species are most likely to occur when the system is at equilibrium.

Now if a living organism that can metabolise some of the metal species formed is added to the system this equilibrium will be disturbed. The metal-containing species absorbed by the organism will be converted into new species. If these new species are formed in process which involve an input of chemical energy, it is difficult to predict with any degree of certainty the distribution of the metal-containing species in the organism or indeed in any part of the system. It is possible, however, that at a certain point in time a steady state or pseudo-equilibrium may be attained. Nevertheless, it will be difficult to apply the usual thermodynamic approach, without making assumptions that may not necessarily be accurate, for the prediction of the distribution of chemical species in living systems. In such systems the experimental determination of the species is the more reliable method.

One of the risks of using the experimental approach is that the species under investigation may be modified or may not remain stable during the various manipulation and analytical steps. The destruction of a fraction of a species could be tolerated if the investigator is only interested in what I shall call qualitative speciation. With this approach, the aim is to determine what major species are present in the system and in some cases to establish the presence or absence of a particular species. In contrast, quantitative speciation is an approach whereby the relative distribution of the species present is determined. Here, more stringent control of extraneous contamination is required and the loss of the species is not acceptable. When using this approach the recovery of the element after each manipulation step should be given.

It is appropriate at this point to discuss the various approaches used to describe the stability of complexes. In order to be able to detect a species, it must be stable in the time scale of the measurements. In other words, during the experiments other competing metals or ligands present in the medium as contaminants or reagents should not substitute the metal or ligand. Furthermore, the experimental conditions should be such that the species does not spontanously dissociate. Both conditions could be satisfied when the complex has the appropriate thermodynamic and kinetic properties. These properties are determined in part by the experimental conditions.

It is therefore apparent that as an aid in choosing the appropriate techniques and methods used for speciation some knowledge of the stabilities and reactivities of the complexes under investigation should be available. Unfortunately, this information is not always available for various reasons. First, the ligands associated with the metals are not always known. Second, the binding sites of the metals are sometimes not fully characterised. Third, the complexity of some biological molecules introduce difficulties in the measurements of the above parameters. Although this lack of information is certainly a drawback, it is however, possible to make some intelligent guesses on the basis of the knowledge gained from the behaviour of simple metal complexes in solution.

According to Taube [9], complexes can be divided into two classes i.e. inert and labile depending on the rate at which the substitution reaction occurs. The author defines labile complexes: as those which take part in substitution reaction without any delay (~ 1 min) on coming into contact with other reagents under ordinary conditions. These conditions are room temperature and concentration of the reagents of about 0.1 M. In contrast, inert complexes are slower to react. In this article, the author discusses the reasons for the differences in the kinetic behaviour of various inorganic complexes and also attempts to classify them in terms of their lability. Another detailed account on the same topic can be found in the book by Basolo and Pearson [10].

On the basis of the information presented in both sources, it is possible to make some generalisations that are relevant for speciation. First, the rate of substitution is determined by the constituents and nature of the medium. This implies that competing ligands and metal ions have to be absent, and the reagents used should be examined for their reactivity towards the metal under investigation. Second, biological molecules with the ability to form chelates will be relatively inert complexes compared with their monodentate analogs. Third, metal complexes formed by the transition elements are more stable than similar non-transition compounds.

Another important consideration is the thermodynamic stability of the complex. Consider the simplified representation of the interaction between a metal or metal-containing complex M, and a biological molecule or an inorganic ligand L, to form the complex ML given by the following Eq.:

$$M + L \rightleftharpoons ML \tag{1}$$

A measure of the thermodynamic stability of ML is given by the equilibrium stability constant

$$K = \frac{\{ML\}}{\{M\}\{L\}} \tag{2}$$

where the brackets { } denote the activity of the individual species. Note that the arrows indicate that the system is at equilibrium, and the charges on the species have been omitted. The interaction between M and L could involve either covalent, ionic or other binding forces.

A more common form of Eq. (2) is given by the equilibrium quotient

$$K' = \frac{[ML]}{[M][L]} \tag{3}$$

Eq. (3) can be converted into Eq. (2) by the following transformation

$$K = \frac{[ML]\,\gamma_M}{[M]\,\gamma_M[L]\,\gamma_L} = \frac{\{ML\}}{\{M\}\{L\}} = K'\frac{\gamma_{ML}}{\gamma_M\,\gamma_L} \tag{4}$$

where γ_{ML}, γ_M and γ_L are the activity coefficients of ML, M, and L, respectively. The stability constant K is related to the free energy changes in a reaction by the following Eq.:

$$-\Delta G^\theta = RT \ln K \tag{5}$$

where ΔG^θ is the change in standard free energy, R is the gas constant and T the absolute temperature. The more negative the value of ΔG^θ, the more likely is the reaction to proceed to the right. The formation of ML is favoured when the value of K is high or in other words, the dissociation of ML into its component constituents, M and L, is not favoured.

The factors that determine the value of the stability constant depend on the nature of the metal and ligand. Attempts at introducing a systematic approach in predicting the stability of complexes have shown that certain ligand atoms prefer to bind particular metal ions. The various concepts for the rationalisation of this preference have been developed by Sidgwick [11], Ahrland et al. [12], and more recently Pearson [13,14]. Although these concepts contain some unresolved contradictions, they are nonetheless useful in giving an idea of the type of complex expected and the possible sequence of the stability.

In general, the transition metal ions form complexes of higher thermodynamic stability compared to the alkali and alkaline earth metals. This is borne out by comparing the roles of the elements Ca, Mg, K and Na with those of Cu and Fe in biological systems. The former group of elements are mobile or semimobile and one of their main functions is as charge carriers. In contrast, the latter group of elements are tightly bound. Although, the interaction between the alkali metals and most ligands are predominantly ionic in nature, it will be an oversimplification to suppose that this is the overriding factor that influence their stability. The architecture of the complex is also important. The implication of the above discussion for speciation is that due to their thermodynamic and kinetic stability, transition metal complexes could be fractionated by a wide variety of techniques. This classification is not restrictive and other stable non-transition metal complexes may also be similarly separated.

A useful concept for the classification of metal-containing proteins has been suggested by Vallee [15]. The proteins are divided into two groups: metalloproteins and metal-protein complexes, on the basis of their stability during the isolation procedures. Metalloproteins retain their metal constituent during fractionation and there is a stoichiometric relationship between the metal and protein. On the other hand, the metal is loosely bound and easily lost during dialysis in metal-proteins. Examples of both types of proteins can be found in an article by Vallee and Coleman [16].

In biological systems metals and metalloids not only interact with high molecular mass constituents, for example proteins and DNA, but also with a host of low molecular mass ligands, amino acids, peptides, inorganic ligands and others. The above discussion also apply to these species. A fraction of some elements are unbound and these could be treated as hydrated free ions.

Although, some species may tolerate a wide range of experimental conditions, however, extreme conditions could cause their destruction. Conditions that affect the speciation of a metal are discussed below.

3.1 Experimental Conditions that Affect the Metal-ligand Interactions

All constituents present in a system interact with each other to a certain extent. The nature and extent of the interaction is determined in part by the nature of the species involved and the properties of the medium in which the experiments occur. Some of these properties include the pH, ionic strength, ionic composition, temperature and dielectric constant. In addition, some species are very unstable in the presence of ultraviolet light and oxygen.

3.1.1 The Effect of Changes in Dielectric Constant

For species that are involved in electrostatic interaction a change in the dielectric constant of the medium will alter the extent of the interaction. The force of interaction between the charged species is inversely proportional to the dielectric constant, assuming that other factors are kept constant. Therefore on substituting a medium of high dielectric constant, for example water, for one of a lower value, most organic solvents, the forces of attraction between the species will increase. As a result there is a greater tendency for association and formation of ion-pairs.

Changes in the dielectric constant could result in the aggregation of large molecules and these could be precipitated out of the solution.

3.1.2 The Effect of Ionic Strength

The activity coefficient of a species is related to the ionic strength by the Debye-Hückel Eq.

$$\log \gamma = \frac{-A_\gamma Z^2 I^{1/2}}{1 + B_\gamma a I^{1/2}} \tag{6}$$

where I is the ionic strength of the solution, Z is the charge on the species, a is the distance of closest approach and A_γ and B_γ are constants which involve numerical factors, temperature and dielectric constant. As can be seen from Eq. 6, variation in activity coefficient occurs with changes in ionic strength. As a consequence, the stability constant of the metal complexes would change (see Eq. 4).

The effect of increasing the ionic strength is that the interionic distances are reduced and the interactions between charged species in the medium are increased. This could result in the precipitation of some species.

3.1.3 The Effect of pH and Temperature

Extreme changes in pH and temperature result in the denaturation of proteins and as a consequence some enzymes may be inactivated. A review by Dawes [17)] covers this topic.

An example of the effect of these two factors on protein denaturation is given in the paper by Levy and Benaglia [18)].

As the concentration, or more accurately the activity, of the hydrogen ions increases, they can compete effectively with metals for the ligand binding sites. On the other hand, ligands may be displaced from the metal when the activity of the hydroxyl ions increases.

3.1.4 The Effect of UV Light and Oxygen

Some reactions are catalysed or induced by ultraviolet light and oxygen. The effect of this is twofold: (i) some species will spontaneously dissociate and (ii) new species may be formed. It is therefore clear that wrong results could be obtained.

In order to reduce this effect, the samples should be protected from sunlight by wrapping the sample bottles in appropriate material. When dealing with oxygen sensitive compounds, the reactions should be performed under nitrogen or other inert gas atmosphere.

4 Fractionation Techniques

Ideally, in any speciation experiments the goal should be to detect the species which contain the element under study and to fully characterise them. However, most investigators stop short of fulfilling this latter requirement for various reasons. First, a wide range of techniques are required in order to fully characterise most species and the task becomes more difficult when dealing with complex molecules. Second, the amount of isolated material is in most cases limited, in the microgram range or less. Third, it is difficult to achieve the degree of purity which is required in order to unambiguously identify a species. Fourth, the application of some of the techniques is tedious and time-consuming. Finally, the metal-ligand interaction is more likely to be disrupted as the number of analytical steps increases.

In spite of these limitations it is possible to obtain valuable data provided specific questions are posed. For example, it has been suggested that the level of methylmercury in a sample is a good indicator of the level of mercury toxicity [19]. Therefore it will be most useful to detect and quantify this species rather than to use the available resources to detect a wide range of mercury species, whose biological behaviour are yet to be established. It is essential to clearly define the purpose and use to which the information will be put before any speciation experiments are performed. This helps not only in the choice of the appropriate techniques and methods, but also helps to focus attention on the parameters that are useful for the interpretation of the results.

In order to minimise the number of analytical steps, and as a result reduce the probability of disrupting the metal-ligand interaction and/or the introduction of contamination into the sample, it would be advantageous to be able to detect and determine the various species in situ. However, techniques like anodic stripping voltammetry (ASV), fluorescence, electron paramagnetic resonance spectrometry (EPR), ion-selective electrode potentiometry and hydrogen-ion potentiometry that could be used to obtain the necessary information suffer from various drawbacks when applied to biological problems. First, the techniques do not, in some cases, have the required sensitivity. Second, in cases where sensitivity is adequate the above techniques still cannot be successfully applied because of interferences caused by matrix components. Because of these limitations one or more fractionation steps are required in order to separate the various species before the detection stage.

Among the fractionation techniques that have been used to study the speciation of trace element containing species include liquid chromatography, gas chromatography, ultrafiltration, dialysis, protein precipitation, electrophoresis and others. In the following sections the application of the above techniques will be discussed.

4.1 Liquid Chromatography

Over the last few years, developments in instrumentation and column technology have made liquid chromatography a versatile technique that can be used for the fractionation of a wide range of biological molecules. These developments have led to reduction in separation times, improvement in column efficiency and reduction in sample requirements. However, the chances of introducing trace metal contamination into the sample have increased because of the use of more metal components in the equipments. Furthermore, it has been suggested that labile metal-ligand species could dissociate on coming into contact with the metal components [20]. Some of the problems associated with the use of liquid chromatography for speciation have been discussed by Gardiner and Delves [21].

Separation can now be achieved by using any one and/or a combination of the various modes available. These modes include size exclusion, normal phase, reversed-phase, paired-ion reversed-phase and ion-exchange. The principles, theories and the instrumentation of the above techniques can be found in various books [22–25] and articles [26–28]. Although liquid chromatography in all its modes of separation has been applied to the fractionation of a range of biological molecules, few examples exist of its use in the area of chemical speciation. Therefore in the discussions below some attention will be given to examples that can be applied after slight modification to the study of various metal-containing species in biological materials. Readers seeking articles on the application of chromatography for the fractionation of metal complexes are refered to the articles by Schwedt [29], Nickless [30], Krull [31], Veering and Willeford [32], and Cassidy [157].

4.1.1 Size Exclusion Chromatography

Separation by size exclusion chromatography is achieved by exploiting the differences in the molecular size. This mode of separation offers a number of advantages when applied to the speciation of metal-containing species in biological materials. First, the pH and ionic strength of the eluant can be chosen to correspond with that found in the sample. This is especially useful when fractionating species with weakly bound metal constituents. Second, interference effects caused by the non-ideal behaviour of column packing material are minimum. These effects can adversely affect the resolution of the method and could lead to the disruption of the metal-ligand association. Third, most samples can be directly applied to the columns with the minimum of sample pretreatment. Finally, the operating conditions are compatible with those of various element-selective detectors. Therefore, it is possible to carry out the fractionation and detection steps simultaneously. The above reasons make size exclusion chromatography an attractive technique to apply in the first instance when studying an unknown sample.

Barth [33] has examined the various aspects of size exclusion chromatography (SEC) relevant for biological samples and these include
(i) a list of commercially available column packing materials,
(ii) column calibration techniques, and
(iii) the various non-size exclusion effects
that could interfere with the resolution of the technique.

Size exclusion chromatography also known as gel filtration has been extensively used as a part of the isolation protocol of various biological molecules. This mode of application will not be considered, instead this article will concentrate on examples that use SEC as the only fractionation technique in combination with detection methods that help to give an indication of the type of molecules associated with the metal.

One of the first attempts at using SEC to study the distribution of trace metals was by Fritze and Robertson [34]. They used gel filtration, ultraviolet detection and instrumental neutron activation to study the distribution of Cu, Fe, Zn, Al and Mn in serum. A subsequent paper [35] from the same laboratory dealing this time with only copper highlighted the contamination problem that could occur on the column, and how this could lead to errors in the identification of the metal complexes. Two important observations made by the authors are noteworthy,
(i) the shape of the metal distribution after the fractionation on the column is almost Gaussian shaped,

(ii) tracer experiments could be used to help distinguish between biosynthesised metal-containing proteins from those formed by in-vitro contamination of the sample.

Subsequent work by Gardiner et al. [36] showed that in a relatively complex mixture like human serum, the association between a metal and protein or ligand could be said to have been established when their Gaussian distribution coincide. This is more likely to be true if the elution volumes of the constituents are in the fractionation range rather than in the excluded volume. Examples of the usefulness of immunological techniques as an aid in identification of proteins are given. Necessary clean-up procedures are also suggested.

An interesting application of speciation is in the study of changes in the distribution of certain metals after administration of drugs. Falchuk [37] used gel filtration combined with flame atomic absorption spectrometry to study effects of the administration of ACTH on the zinc distribution in serum. Kamel et al. [38] developed methods to follow the distribution of gold.

Human serum is relatively more characterised than other biological samples and as a result some information on various metal binding proteins exists in various reference books. However, in samples for which such information is not available most investigators have resorted to calibrating the column with proteins or molecules of known molecular masses. Sample types to which this approach has been applied include serum, milk, amniotic fluid, urine and tissue homogenates. Some of the applications of SEC are summarised in Table 1.

Because of the complex nature of most biological samples, a single fractionation technique may not be adequate for the separation of the wide range of molecules present. Better resolution of some molecules is obtained when properties other than differences in size are exploited. These include differences in ionic characteristics, affinity for other molecules and hydrophobicity. In separations that involve any one or more of these properties, the sample constituents interact with the column material and are then eluted with a suitable eluant. As a consequence of this interaction, and the use of eluants, whose properties may not closely resemble those of the medium found in vivo, the metal may dissociate from the ligand. In addition, as the complexity of the sample increases it is difficult to predict the behaviour of the various constituents. Undesirable effects leading to irreversible interaction between some molecules in the sample and the column packing material, degradation and decomposition of some constituents may result. Furthermore, it may be difficult to rid the column of certain trace metal contamination.

Some of the above mentioned problems can be reduced by applying pretreated samples to the column. The samples may be partially digested or passed through a precolumn [51]. It is therefore essential to provide information showing that the metal-ligand interaction is not disrupted during the pre-separation steps and no change in the state of the metal has occured.

4.1.2 Ion-exchange Chromatography (IEC)

Species with different ionic characteristics interact to varying degrees with a suitable column packing material under given experimental conditions. This difference in behaviour forms the basis of separation by IEC.

Table 1. Selected examples of the application of size exclusion chromatography

Matrix	Elements Examined	Other Methods and Techniques used together with SEC	Comments	Ref.
Amniotic Fluid	Cu, Fe and Zn	GFAAS and Immunonephelometry	Identification of proteins associated with the metals by immunonephelometry	[39]
Milk	Zn	UF, IEC, IR, NMR, Citrate and Prostaglandin assays	Example of the use of a wide range of techniques that are required for species detection and identification	[40]
Serum	Br, Cu, Fe, I, Mn, Se, Rb, V and Zn	Neutron activation	Detailed examination of the behaviour of the elements during fractionation and the effect of contamination in the reagents and equipments	[41]
	Cu, Zn	GFAAS, Immunonephelometry		[36]
	Zn	GFAAS, Affinity chromatography, Kinetic immunoturbidimetry		[42]
	Al	GFAAS, Ultrafiltration		[43]
Synthetic Mixture of Proteins and Enzymes	Cu, Fe, Mn, P, Zn	ICP		[44]
Tissue Homogenates				
Liver	As, Cd, Cu, Hg, Se, Zn	Neutron activation	This investigation demonstrate that after gentle sample pretreatment the protein-bound trace element may remain intact	[45]
Liver and the digestive of the Oyster System	Cd, Cu, Zn	FAAS, GFAAS, Ultracentrifugation	The methods chosen could be used for the study of the distribution of trace element containing species in subcellular particles	[46]
Kidney and the digestive system of the horse mussel	Cd, Cu, Fe, Pb, Mg and Mn	FAAS, GFAAS, Ultracentrifugation		[47]
Lumen of the small intestine	Pb	Ultracentrifugation, Gamma counting		[48]
Lobster digestive gland extracts	Cd, Cu, Zn	FAAS		[49]
Renal cytosol and microsomes	Ni	Ultracentrifugation, Liquid scintillation counting		[50]

Table 2. Selected examples of the application of ion-exchange chromatography

Matrix	Elements Examined	Additionally Methods and Techniques used	Comments	Ref.
Plasma and Urine	As	X-ray Spectrometry Gamma counting	Fractionation of arsenic metabolites formed after in vivo incorporation experiments with ^{74}As	53)
Marine Organisms	As	Homogenisation, Extraction and GFAAS		54)
Plant Material	As	Extraction and Hydride generation	Separation of arsenite and arsenate	55)
Plant Material	As	Extraction and Hydride generation		56)
Tissue	As	Extraction, Digestion and Hydride generation	Differentiation of As(III) and As(V)	57)
Urine and Blood	As	Digestion and Hydride generation		58)
Fish Tissue	As	Extraction, IR and TLC		59)
Urine	As		Both inorganic and organic Arsenic compounds could be identified	60)
Biological samples	As	ICP		61)
Urine	Cr	Extraction, ICP	Differentiation of Cr(VI) and Cr(III)	62)
Urine	Cr	Extraction, GFAAS	Differentiation of Cr(VI) and Cr(III)	63)
Blood	Cr	Extraction, GFAAS	Differentiation of Cr(IV) and Cr(III)	64)

The experimental parameters that could be manipulated in order to achieve a separation include the pH, temperature and ionic strength of the mobile phase. As already discussed these factors can affect the metal-ligand interaction. Another complicating factor is the use of a charged surface for the separation. It is therefore to be expected that labile metal-containing species may not remain intact on applying this technique. In fact, one of the methods of measuring the proportion of labile to inert species in a sample uses IEC. The sample is passed through a column that will bind the free ions and labile species, and the compounds that do not interact with column packing material are regarded as kinetically stable.

Various column packing materials and the experimental conditions under which they can be properly used have been summarised by Rabel [52].

Although IEC has been used to separate large molecules most application have been in the fractionation of low molecular mass species. Some of the applications to the study of the speciation of trace elements in biological materials and the necessary pretreatment steps are summarised in Table 2.

4.1.3 Reversed Phase Chromatography (RP)

Separation by RP is achieved by exploiting the differences in the hydrophobic characteristics of the molecules. Recent developments in column packing materials have resulted in the increased application of RP for the fractionation of biological molecules. A book by Krstulović and Brown [23], and a number of articles [65-68] have dealt with various aspects of this technique.

Applications that could be of relevance to species identification in biological materials can be found in various reviews [31,69,70]

4.2 Gas Chromatography (GC)

Although most biological materials that are not already in the liquid form can be readily homogenised and converted into a form which is suitable for use in liquid chromatography, however, better resolution of some constituents are obtained when they are first extracted and then injected into a gas chromatograph. In fact in cases, where the species are volatile at low temperatures sample pretreatment may not be necessary. The vapour phase that is in equilibrium with the sample in a closed system could be directly injected into a gas chromatograph using the techniques of headspace analysis [71]. Another possibility of using GC for speciation in solid samples has been suggested by Bächmann [72]. Furthermore, GC is well suited for the study of speciation in breath and expired air, an area of research that may be of interest in the future.

Species that are fractionated by GC have to fulfil some requirements. First, they have to be volatile or readily converted into volatile forms. Second, the species have to be thermally stable at the operating temperatures of GC columns. Third, degradation of the compounds during the separation step should be negligible.

The above requirements limit the number of species that can be fractionated by this technique. However, hydride forming elements and those that can be converted into volatile alkyl derivatives can be used. Lead [73], tin [74], arsenic [75], mercury [76,77], selenium [78], antimony [88], germanium [88], and thallium species can be determined using the above approach.

Table 3. Selected examples of the application of gas chromatography

Matrix	Elements Examined	Additional Methods and Techniques used	Comments	Ref.
Urine and Sweat	Hg	GFAAS	No sample pretreatment is required	79)
Animal tissue	Hg	Cold-vapour AAS, Isotope dilution analysis		80)
Rat organs	Hg	Extraction		81)
Fish	Hg	Mass spectrometry, Activation analysis, Extraction	identification and determination of methylmercury compounds	90)
Fish	Pb	Extraction, AAS	Detection of five R_4Pb type species	82)
Fish	Pb	Extraction, AAS	Speciation of alkyllead components, both molecular and ionic	83)
Biological materials	Pb	Extraction, GFAAS		84)
Biological samples	Se	Extraction, Digestion	Both organoselenium and selenite(SeIV) can be differentiated by the method developed	85)
Urine	As	Multiple ion detection, Hydride generation heptane cold trap, Extraction	Inorganic, monomethyl-, dimethyl- and trimethylarsenic compounds were detected by this combination of techniques	86)
Urine	As	Extraction, Flame-photometric detection	Species detected include inorganic, As(III) and monomethylarsonic acid	87)
Cell Culture	As	Mass spectrometry	Identification of arsenate, arsenite, monomethyl-arsonate and dimethylarsinate	89)

An advantage of using GC is that it can be easily coupled with mass spectrometry, a technique that can be used to elucidate the structure of the species.

The application of GC for the separation of metal complexes has been reviewed by various authors [30,105]. Some of the methods that have been applied to biological materials are summarised in Table 3.

4.3 Electrophoresis and Related Techniques

Electrophoresis is one of many electromigrational separation techniques [91] which include isotachophoresis, immunoelectrophoresis and isoelectric focussing that have been used to separate various species on the basis of their different mobility in an electric field. These techniques can be used not only to achieve separations but also it is possible to identify the ligand bound to the metal. This can be done by comparing the isoelectric points, immunological behaviours, extent of mobilities or step heights of the sample constituents with those of well-characterised standards. A difficulty, however, is in the determination of the metal constituent itself. Except in the case of radioisotopes, the activities of which can be easily measured, non-radioactive elements can be detected only after further separation steps.

Because of the use of various electrolyte systems, pH gradients, and not least an electric field, some complexes would not survive the separation. It is therefore necessary that the species to be separated are both thermodynamically and kinetically stable. Recently, Boček and Foret [92] have reviewed the application of isotachophoresis to the separation of inorganic species. This technique appears to be well-suited for the study of the distribution of metabolites of metal-containing drugs in body fluids. A survey of the application of electrophoretic techniques to biological materials can be found in the book edited by Deyl [93].

4.4 Ultrafiltration and Dialysis

Ultrafiltration and dialysis are separation techniques that have been extensively used for the removal of low molecular mass constituents from biological fluids. Separation is achieved by forcing the molecules through filters or membranes with the appropriate pore sizes. The necessary force is generated either by a pressure gradient, ultrafiltration, or concentration gradient, dialysis. Kwong [94] in a recent review of the measurement of free drug levels in body fluids, has discussed both techniques in detail. Most of the comments regarding the application of both techniques are of relevance for the speciation of metal-containing species.

Although the resolution of the constituents achieved by ultrafiltration and dialysis are lower than those of either liquid chromatography or electrophoresis, they could be used to obtain rapid separations in cases where it is known that there are two major fractions with widely different molecular mass.

Since the quality of the separation is determined by the properties of the filter used, it is essential that the investigator should understand the causes of their non-ideal behaviour and how these can be minimised. The non-ideal behaviour is due to nonspecific adsorption of the constituents to be separated on the filter, Donnan equilibria, leakage of high molecular mass constituents and hindered passage of low molecular mass species through the filter. Some of these effects and how they influence the speciation results have been described by Gardiner and Delves [21].

5 Detection of Species

As already stated, complete speciation involves the identification, determination and characterisation of both metals and ligand constituents of the species. Although metals and ligands are determined by applying two different types of detector systems, it is essential that the results should clearly show the relationship between the metal(s) and ligand(s) that directly interact. The presence of a metal and a ligand in a fraction is not sufficient evidence for the existence of direct interaction. Ideally, the detector should respond specifically to the presence of the whole species. However, very few species can be detected in this manner.

The detection of metals and metalloids no longer present major analytical problems. The instrumental techniques are both sensitive and specific for most elements. In contrast, the usual techniques for the detection of biological molecules respond to functional groups and consequently, they are relatively non-specific. However, it is possible to apply more specific methods when some information is available about the likely identity of the molecule.

Both types of detector systems can be used in two modes of applications, i.e., on-line and off-line. In the on-line mode the fractionation and detection systems are directly coupled. It is possible to use more than one detector either in series or in parallel. Examples of this mode of detection will be given in the section dealing with combined techniques. The off-line mode involves the collection of the fractions with subsequent determination of the constituents. The advantage of this approach is that further sample pretreatment procedures could be applied where necessary before the constituents are detected. In addition, quantitative estimates of the recoveries can be made.

The techniques that can be used for the determination, identification and characterisation of species in biological materials are discussed below.

5.1 Determination of Metals and Metalloids

The concentration levels of most trace metals and metalloids lie below 1000 $\mu g\ l^{-1}$. Therefore, the classical methods of analysis do not have the required sensitivity. Among the instrumental techniques that have been extensively used for the analysis of biological materials include, atomic absorption spectrometry, plasma emission spectrometry, anodic stripping voltammetry and neutron activation analysis.

5.1.1 Atomic Absorption Spectrometry

Both flame and graphite furnace atomic absorption spectrometry are two of the commonest techniques used for the determination of metals and metalloids. Various authors [95–98] have discussed the application of both to the analysis of trace elements in biological materials.

Flame atomic absorption spectrometry (FAAS) can be used to detect most elements present at levels greater than about 100 $\mu g\ l^{-1}$. For more sensitive determinations graphite furnace atomic absorption spectrometry (GFAAS) is the technique of choice. In addition, if the volume of the fraction is limited GFAAS is ideally suited for the determination because only a few microlitres (5–20 μl) of sample is usually required for an analysis. This is in contrast to FAAS for which a minimum of about 100 μl is desirable.

Determinations by both techniques can be subject to chemical and/or physical interference effects caused by the sample matrix. However, after fractionation of the sample the species are usually in a less complex matrix, a buffer or electrolyte solution. Consequently, matrix interferences effects are minimised. On the other hand, the species may be diluted in the process and this could be detrimental for the determination of species present at very low concentrations. At the present state of the art GFAAS can be used for the determination of analytes at the 1 $\mu g\, l^{-1}$ level. However, at this level contamination in the reagents and equipment limit the number of species that can be detected with confidence.

In some cases, it is essential to use a sensitive technique like GFAAS in order to measure the level of contamination in the reagents used. This is important because the presence of extraneous contamination may lead to the formation of artifacts or the amount of element associated with particular species (those that can incorporate the analyte in-vitro) may be overestimated.

The application of atomic spectroscopic instruments as element-specific detectors in chromatography has been reviewed by van Loon [99]. More recently, Krull [31] has extensively reviewed their use in high pressure liquid chromatography (HPLC). Atomic spectrometry has found wide acceptance in the field of liquid chromatography because, in most cases, the fractions can be directly analysed after elution from the column. However, it is possible to use the technique for the analysis of solid samples without first dissolving the matrix. This is particularly useful after electrophoresis, where the fractions are fixed either in a gel or on paper. Kamel et al. [38] have shown that it is possible to cut the appropriate sections and insert them into the carbon furnace for analysis. The disadvantage of this approach is that the precision is usually poorer (about 10%) and it is difficult to calibrate the instrument. Nevertheless, this approach is very useful if it is used for qualitative speciation.

Hydride forming elements, Se, As, Hg, have been determined by transfering the hydride evolved after chemical pretreatment of the sample into a flame or onto a carbon furnace. Several authors have reviewed the application of both technique to the analysis and speciation [75,88] of Se, As, Hg, and Pb. Attempts at using hydride formation as a way of achieving speciation without prior fractionation of the species have not been very successful, because of the difference in behaviour between the organic and inorganic forms of these elements. Consequently, for the present, attempts to differentiate the various oxidation states and species of these elements require a chromatographic step or other form of pretreatment before detection.

Recently, Sakai et al. [100] have combined flame Zeeman atomic absorption spectrometry (FZAAS) with selective vapourisation of the species from a sample, placed in a crucible which is slowly heated, to investigate the speciation of arsenic compounds in oyster tissue. This method could prove useful if the top temperature reached by the system is high enough to allow the vapourisation of a wider variety of species that may exist in biological samples. Presently, the highest temperature attainable is 400 °C.

5.1.2 Plasma Emission Spectrometry

Among the plasma sources that have been used for analytical measurements include the inductively coupled argon plasma (ICP), direct current argon plasma (DCP) and microwave induced helium plasma (MIP). The instrumentation and performance of the more popular ICP source have been discussed by Barnes [101]. More recently, Thompson and Walsh [131] have published a book dealing with the practical aspects of ICP.

The advantages of using plasma emission sources include the ability to perform multi-element analysis, a calibration linear dynamic range of more than three orders of magnitude and for some elements the limits of detection are comparable to those found by GFAAS. The ability to perform multi-element analysis is essential when the purpose of the experiments is to study element interaction effects.

Some authors [102,103] have found that the sensitivity of the determination is influenced by the oxidation state of the element and the molecular form. This should be borne in mind when quantitative speciation is contemplated.

Besides liquid samples, gases and solids [104] can be analysed after making the appropriate modifications to the sample introduction system. The application of plasma sources as detectors for gas chromatography of metal complexes have been reviewed by Uden [105]. Literature dealing with the analysis of gas and liquid chromatographic effluents have been surveyed by Carnahan et al. [106].

5.1.3 Stripping Voltammetry

Stripping voltammetry is an electrochemical technique which can be used to detect electroactive species. This process occurs in two steps. The species are either reduced or oxidised, depending on the nature of the complex, onto a suitable electrode at a given fixed potential. This is followed by stripping out the constituents on the electrode by applying a scanning voltage. The characteristic current generated on the electrolysis of each species appear as peaks in the voltammogram. If the oxidation or reduction potentials of the species are sufficiently separated, each peak can be related to a specific complex.

In a simple solution containing a background electrolyte, a metal ion and a ligand, stripping voltammetry (SV) can provide information on the stability constants and kinetic labilities of the complexes formed. Furthermore the concentration of the free metal ion and ligand could be determined; it can also be used to distinguish between different oxidation states of the metal ion. However, in the presence of a complex matrix some of this information is difficult to obtain. The adsorption of protein or other molecules on the surface of the electrodes could hinder the electrochemical process and thus interfere with the determinations. Therefore SV can only be applied to complex biological matrices after sample digestion or some other appropriate treatment. Nevertheless, if the species can be separated from the matrix intact, the successful application of SV could provide the type of information that could lead not only to the identification and characterisation of the species but also provide the basis for predicting its biochemical behaviour.

The principles, instrumentation and applications of stripping analysis can be found in a number of publications [107-110]. Cammann [111] has recently reviewed the application of electrochemical techniques for the speciation of anions.

5.1.4 Neutron Activation Analysis (NAA)

Unstable radionuclei result on subjecting the nuclei of some elements to neutron bombardment. During the decay process, in which the radionuclei return to more stable forms, characteristic radiation is emitted. The energy of the radiation is characteristic of the element, and its intensity forms the basis for quantitative elemental analysis. The advantages of NAA for trace analysis include low detection limits, good sensitivity, multi-element capability and relative freedom from matrix effects. However, for successful application of this technique skilled personel are required and because of the low sample throughput the amount of work involved in the analysis of column fractions, for example, is prohibitively high. In addition, it may take up to several weeks before the results are available. Further, only few laboratories have easy access to a neutron source.

Recently, Heydorn [1] has dealt extensively with the various aspects of the application of NAA to the analysis of biological materials. The usefulness of neutron activation analysis for the determination of protein-bound elements in human serum has been demonstrated by Woittiez [41].

5.2 Identification of Macromolecular Ligands and Anions

Although metal(s) and metalloid(s) may form a part of the active centre in a species, the biochemical behaviour of the species is determined in part by the mutual effect of the ligand on the metal and vice-versa. As has already been demonstrated in the foregoing discussion, the determination of the metal constituent does not present major difficulties. In contrast, the identification of the ligands is not that straightforward. It is perhaps reasonable to treat the ligand as a single entity rather than as multi-atom constituent. This approach is advantageous because elemental analysis of a complex molecule, for example a protein, will provide the percentage of carbon, nitrogen, sulphur, oxygen and/or any other elements present, however, it may still be difficult to identify the molecule. In addition, it is difficult to obtain the species in a high degree of purity therefore the values for the above mentioned elements may be in error.

Most conventional techniques for the determination of biological molecules or other species with similar properties use their ability to absorb ultraviolet or visible light, their fluorescence after excitation with light of the appropriate wavelength, or their electrochemical behaviour. It possible to enhance the detectability of some species by making them react with UV-visible absorbing or fluorescent compounds. Applied to complex matrices, these detection methods are at best only selective, because a wide variety of chromophores will give a response.

The application of the above techniques to the detection of biological molecules can be found in various reviews [112-117]. In order to identify the molecule, additional information is required. This type of information may be the isoelectric point, an estimate of the molecular mass, or the chemical behaviour. It could then be possible to narrow the number of candidates. At this point, if the molecule is of biological origin, an immunological technique [118-120] could be applied.

Small ligands of both inorganic or organic origin can be determined by ion chromatography coupled with a suitable detection system [121]. Species like AsO_3^{3-}, AsO_4^{3-}, CrO_4^{2-} can be detected by this technique. The principles and application of this technique can be found in books by Fritz et al. [122], and Smith and Chang [123], and a review article by Small [158].

5.3 Characterisation of the Species

It may be difficult to completely identify a species, but nevertheless some information that is of importance to the rationalisation of its biochemical behaviour could be gained. For example, the oxidation state of the metal, environment of the metal, stability and reactivity of the species. The techniques that can be used to obtain this type of information are summarised in Table 4.

Table 4. Techniques that could be used for the characterisation of species

Technique	Information that could be acquired by the technique	Ref.
Mössbauer spectroscopy	Oxidation and spin states of the metal in the prosthetic group. Electronic structure and spatial arrangement of the active centre	124, 125)
Electron spin resonance	Structural and kinetic information on the species	126)
X-ray spectrometric analysis	Information on the structure of the species	127)
Fluorescence spectroscopy	Binding of substrates, association reactions between species, denaturation of proteins and other macromolecules	128)
Circular dichroism	Secondary structure of proteins, interaction between ligands and proteins, binding of metals at active sites in enzymes	129)
NMR	The structure, equilibria and kinetics of the species	130)
Raman and IR	Conformational changes and structural analysis of the species	129, 132)
Electronic spectroscopy	The structure around the metal ion	133)

6 Combined Techniques

Over the last decade developments in instrument automation and data processing have led to the design of more efficient analytical instruments. In a bid to further maximise the amount of information obtained from analytical processing of a sample, various combinations of two or more compatible analytical instruments are being examined. The advantages of using such systems are: the more efficient use of the sample material, time-saving and minimum sample manipulation. Although some of these systems could provide valuable information, they may be too expensive for the small research laboratory.

A summary of various combined techniques and the type of information that they could provide in relation to speciation are summarised in Table 5.

Table 5. Coupled techniques and the information that they provide in relation to speciation

Examples of Coupled Techniques	Type of Information	Ref.
GC — AAS	Metal and metalloid content in separated fractions.	[74]
— ICP	Identification of the species is possible if well-characterised compounds with similar retention	[105, 134]
LC — AFS	times as the constituents in the sample are available.	[99]
— ASV		[135]
— DCP		[136]
— FAAS		[31, 114, 115]
— GFAAS		[137]
— ICP		[114, 115]
GC — FTIR	Identification and structural elucidation of the species.	[138]
— FTIR-MS		[139]
— IR-MS		[140]
— MS		[141, 142]
LC — IR		[140]
— FTIR		[143]
— MS		[144]
— MS-MS		[145, 146]
ICP — MS	Determination of metal content and isotope dilution analysis	[147]
LC — CD	Structural information and identification.	[148]
NMR		[24]
— ESR		[115]
IC — AAS	Determination and identification of anions.	[149]

7 Species Identification with the Aid of Computer Programs

Although the emphasis in this article has been on the discussion of techniques and methods that can be used in the laboratory for the identification of species, increasing importance is being attached to computer simulation of trace element speciation. The reason for this increased interest could be attributed in part to the availability of relevant experimental data which could be used in developing the required models. However, computer simulation comes into its own when the species are so unstable that separation techniques cannot be applied and/or the detection systems do not have the required sensitivity.

The approach used in developing mathematical models for chemical speciation involves using appropriate computer programs to identify the predominate species assuming that the system is at equilibrium. The data required include the concentration of the metal ions and ligands, the stability constants of the probable complexes, and the pH, temperature and ionic strength of the medium. In its most primitive form the computer program will provide information only on 1 : 1 complexes between the metals and ligands. At a more sophisticated level polynuclear and mixed ligand complexes are considered. Further the effect of the presence of a solid or gas phase in contact with the sample and how this affects the complex equilibria is also taken into account.

There are currently at least five major programs availiable each incorporating an aspect of complex equilibria which is judged to be of importance by the authors. The reasoning behind the development of some of these programs and their application to biological systems can be found in articles by Perrin et al. [150–152] and Williams et al. [153, 154].

Computer simulation has been used to predict the speciation of various trace elements during chelate therapy [155] and in total parenteral nutrition [156].

8 Future Trends

A multidisciplinary approach is required in order to achieve total speciation. This approach is not only desirable but essential for the proper design of the experiments and interpretation of the results. Over the next few years, it is to be expected that more information relevant to speciation studies would become available. This would lead to a more informed choice of methods and techniques. In addition, on the basis of this knowledge, accurate computer simulation of the distribution of species in a wider variety of biological systems could be accomplished.

9 List of Abbreviations

AFS	Atomic Fluorescence Spectrometry
ASV	Anodic Stripping Voltammetry
CD	Circular Dichroism
DCP	Direct Current Argon Plasma
EPR	Electron Paramagnetic Resonance
ESR	Electron Spin Resonance
FAAS	Flame Atomic Absorption Spectrometry
FTIR	Fourier Transform Infrared
FZAAS	Flame Zeeman Atomic Absorption Spectrometry
GC	Gas Chromatography
GFAAS	Graphite Furnace Atomic Absorption Spectrometry
HPLC	High Pressure Liquid Chromatography
ICP	Inductively Coupled Argon Plasma
IEC	Ion-Exchange Chromatography
IR	Infrared Spectrometry
MIP	Microwave Induced Helium Plasma
MS	Mass Spectrometry
NAA	Neutron Activation Analysis
NMR	Nuclear Magnetic Resonance
RP	Reversed-Phase Chromatography
SEC	Size Exclusion Chromatography.

10 Acknowledgement

The author will like to thank Dr. M. Stoeppler for his interest, and Dr. D. J. Halls for his useful comments.

11 References

1. Heydorn, K.: Neutron Activation Analysis for Clinical Trace Element Research, Vol. 1, Boca Raton, Florida, CRC Press, Inc. 1984
2. Versieck, J., Barbier, F., Cornelis, R., Hoste, J.: Talanta *29*, 973 (1982)
3. Stoeppler, M.: Analytical aspects of sample collection, sample storage and sample pretreatment, in: Trace Element Analytical Chemistry in Medicine and Biology (ed.) Brätter, P. and Schramel, P., Vol. 2, p. 909, Berlin—New York, Walter de Gruyter & Co 1983
4. Behne, D.: J. Clin. Chem. Clin. Biochem. *19*, 115 (1981)
5. Aitio, A., Järvisalo, J., Stoeppler, M.: Sampling and sample storage, in: Trace Metal Analysis in Biological Specimens (ed.) Stoeppler, M., Foster City, California, Biomedical Publications, in preparation
6. Oda, C. E., Ingle, Jr., J. D.: Anal. Chem. *53*, 2305 (1981)
7. Comprehensive Biochemistry, (ed.) Florkin, M., Stotz, E. H., Amsterdam—London—New York, Elsevier Publishing Company
8. Methods in Enzymology (ed.) Colowick, S. P., Kaplan, N. O., New York—San Francisco—London, Academic Press
9. Taube, H.: Chem Revs. *50*, 69 (1952)
10. Basolo, F., Pearson, R. G.: Mechanisms of Inorganic Reactions, New York, John Wiley & Sons, Inc. 1958
11. Sidgwick, N. V.: The Covalent Link in Chemistry, Ithaca, New York, Cornell University Press 1933
12. Ahrland, S., Chatt, J., Davies, N. R.: Quart. Rev. Chem. Soc. *12*, 265 (1958)
13. Pearson, R. G. : J. Chem. Educ. *45*, 581, 643 (1968)
14. Pearson, R. G.: Surv. Prog. Chem. *5*, 1 (1969)
15. Vallee, B. L.: Adv. Protein Chem. *10*, 317 (1955)
16. Vallee, B. L., Coleman, J. E.: Metal coordination and enzyme action in: Comp. Biochem. (ed.) Florkin, M., Stotz, E. H., Vol. 12, p. 163, Amsterdam—London—New York, Elsevier Publishing Company 1964
17. Dawes, E. A.: Enzyme Kinetics (Optimum pH, Temperature and Activation Energy), in: Comp. Biochem. (ed.) Florkin, M., Stotz, E. H., Vol. 12, p. 87, Amsterdam—London—New York, Elsevier Publishing Company 1964
18. Levy, M., Benaglia, A. E.: J. Biol. Chem. *186*, 829 (1950)
19. Kaiser, G., Tölg, G.: Mercury in: The Handbook of Envirbnmental Chemistry (ed.) Hutzinger, O., Vol 3, Part A, p. Berlin—Heidelberg—New York, Springer Verlag 1980
20. Veening, H., Willeford, B. R.: High-performance liquid chromatography of Metal Complexes, in: Advances in Chromatography (ed.) Giddings, J. C., Grushka, E., Cazes, J., Brown, P. R., Vol. 22, p. 118, New York—Basel, Marcel Dekker, Inc., 1983
21. Gardiner, P. E., Delves, H. T.: The speciation of trace metals and metalloids in biological specimens, in: Trace Metal Analysis in Biological Specimens (ed.) Stoeppler, M., Foster City, California, Biomedical Publications, in preparation
22. Hamilton, R. J., Sewell, P. A.: Introduction to High Performance Liquid Chromatography, London—New York, Chapman and Hall 1982[2]
23. Krstulović, A. M., Brown, P. R.: Reversed-Phase High Performance Liquid Chromatography, New York—Chichester—Brisbane—Toronto—Singapore, John Wiley & Sons 1982
24. Poole, C. F., Schuette, S. A.: Contemporary Practice of Chromatography, Amsterdam—Oxford—New York—Tokyo, Elsevier 1984
25. Parris, N. A.: Instrumental Liquid Chromatography, Journal of Chromatography Library — Vol. 27, Amsterdam—Oxford—New York—Tokyo, Elsevier 1984

26. Hearn, M. T. W.: High-performance liquid chromatography and its application to protein chemistry, in: Advances in Chromatography (ed) Giddings, J. C., Grushka, E., Cazes, J., Brown P. R., Vol. 20, p. 1, New York—Basel, Marcel Dekker, Inc. 1982
27. Regnier, F. E.: Science *222*, 245 (1983)
28. Hearn, T. W., Regnier, F. E., Wehr, C. T.: Intern. Lab., Jan/Feb 1983
29. Schwedt, G.: Topics Curr. Chem. *85*, 159 (1979)
30. Nickless, G.: J. Chromatogr. *313*, 129 (1985)
31. Krull, I. S.: Trace metal analysis by high-performance liquid chromatography, in: Liquid chromatography in Environmental Analysis (ed.) Lawrence, J. F., p. 169, Clifton, New Jersey, The Humana Press, Inc. 1984
32. Willeford, B. R., Veening, H.: J. Chromatogr. *251*, 61 (1982)
33. Barth, H. G.: J. Chromatogr. Sci. *18*, 409 (1980)
34. Fritze, K., Robertson, R.: J. Radioanal. Chem. *1*, 463 (1968)
35. Evans, D. J. R., Fritze, K.: Anal. Chim. Acta *44*, 1 (1969)
36. Gardiner, P. E., Ottaway, J. M., Fell, G. S., Burns, R. R.: Anal. Chim. Acta *124*, 281 (1981)
37. Falchuk, K. H.: N. Engl. J. Med. *296*, 1129 (1977)
38. Kamel, H., Brown, D. H., Ottaway, J. M., Smith, W. E.: Analyst *102*, 645 (1977)
39. Gardiner, P. E. Rösick, E., Rösick, U., Brätter, P., Kynast, G.: Clin. Chim. Acta *120*, 103 (1982)
40. Lönnerdal, B., Stanislowski, A. G., Hurley, L. S.: J. Inorg. Biochem. *12*, 71 (1980)
41. Woittierz, J. R. W.: Elemental Analysis of Human Serum and Human Serum Protein Fractions by Thermal Neutron Activation, Netherlands Energy Research Foundation Report, ECN 147, January 1984
42. Foote, J. W., Delves, H. T.: Analyst *109*, 709 (1984)
43. Leung, F. Y., Hodsman, A. B., Muirhead, N., Henderson, A. R.: Clin. Chem. *31*(1), 20 (1985)
44. Morita, M., Uehiro, T., Fuwa, K.: Anal. Chem. *52*, 349 (1980)
45. Norheim, G., Steinnes, E.: Anal. Chem. *47*, 1688 (1975)
46. Julshamn, K., Andersen, K.-J.: Comp. Biochem. Physiol. *75A*, 9 (1983)
47. Julshamn, K., Andersen, K.-J.: Comp. Biochem. Physiol. 75*A*, 17 (1983)
48. Partridge, S., Blair, J. A., Morton, A. P.: Lead speciation and G. I. adsorption, in: International Conference on Heavy metals in the Environment (ed.) Lekkas, T. D., Vol. 1, p. 423, CEP Consultants Ltd. Edinburgh 1985
49. Guy, R. D., Chou, C. L., Uthe, J. F.: Anal. Chim. Acta *174*, 269 (1985)
50. Sunderman, Jr., F. W., Mangold, B. L. K., Wong, S. H. Y., Shen, S. K., Reid, M. C. Jansson, I.: Res. Commun. Chem. Pathol. Pharmacol. *39*, 477 (1983)
51. de Jong, G. J., Zeeman, J.: Chromatographia *15*(7), 453 (1982)
52. Rabel, F. M.: Ion-exchange packing for HPLC separations care and use, in: Advances in Chromatography, Vol. 17, p. 53, New York—Basel, Marcel Dekker, Inc. 1979
53. Tam, K. H., Charbonneau, S. M., Bryce, F., Lacroix, G.: Anal. Biochem. *86*, 505 (1978)
54. Maher, W. A.: Anal. Chem. Acta *126*, 157 (1981)
55. Austernfeld, F.-A., Berghoff, R. L.: Plant and Soil *64*, 267 (1982)
56. Aggett, J., Kadwani, R.: Analyst *108*, 1495 (1983)
57. Weigert, P., Sappl, A.: Fresenius Z. Anal. Chem. *316*, 306 (1983)
58. Foà, U., Colombi, A., Maroni, M., Buratti, M., Calzaferri, G.: Sci. Total Environ. *34*, 241 (1984)
59. Hanaoka, K., Tagawa, S.: Bull. Japan. Soc. Sci. Fish., *51*(4) 681 (1985)
60. Lovell, M. A., Farmer, J. G.: Human Toxicol. *4*, 203 (1985)
61. Morita, M., Uehiro, T., Fuwa, K.: Anal. Chem. *53*, 1806 (1981)
62. Mianzhi, Z., Barnes, R. M.: Spectrochim. Acta *38B*, 259 (1983)
63. Minoia, C., Mazzucotelli, A., Cavalleri, A., Minganti, V.: Analyst *108*, 481 (1983)
64. Richelmi, P., Baldi, C., Minoia, C.: Intern. J. Environ. Anal. Chem. *17*, 181 (1984)
65. Cooke, N. H. C., Olsen, K.: J. Chromatogr. Sci. *18*, 512 (1980)
66. Bidlingmeyer, B. A.: J. Chromatogr. Sci. *18*, 525 (1980)
67. Karger, B. L., Giese, R. W.: Anal. Chem. *50*, 1048A (1978)
68. Colin, H., Guiochon, L. J.: J. Chromatogr. *141*, 289 (1977)
69. Buckingham, D. A.: J. Chromatogr. *313*, 93 (1984)

70. Cerrai, E., Ghersini, G.: Reversed-phase extraction chromatography in inorganic chemistry, in: Advances in Chromatography (ed.) Giddings, J. C., Grushka, E., Cazes, J., Brown, P. R., Vol. *9* p. 1, New York—Basel, Marcel Dekker, Inc
71. Hachenberg, H., Schmidt, A. P.: Gas Chromatographic Headspace Analysis, London—New York—Rheine, Heyden & Son Ltd. 1977
72. Bächmann, K.: Talanta *29*, 1 (1982)
73. Chau, Y. K., Wong P. T. S.: Organic lead in the aquatic Environment, in: Biological Effects of Organolead Compounds (ed.) Grandjean, P., p. 22, Boca Raton, Florida, CRC Press, Inc. 1984
74. Harrison, R. M., Hewitt, C. N.: Int. J. Environ. Anal. Chem. *21*, 89 (1985)
75. Fishbein, L.: Int. J. Environ. Anal. Chem. *17*, 113 (1984)
76. Zarnegar, P., Mushak, P.: Anal. Chim. Acta *69*, 389 (1974)
77. Rodriguiez, J. A.: Talanta *25*, 299 (1978)
78. Verlinden, M., Deelstra, H., Adriaenssens, E.: Talanta *28*, 637 (1981)
79. Robinson, J. W., Wu, J. C.: Spect. Letters *18*(1), 47 (1985)
80. Harms, U., Luckas, B., Lorenzen, W., Montag, A.: Fresenius Z. Anal. Chem. *316*, 600 (1983)
81. Knoll, R., Woggon H., Macholz, R., Kujawa, M., Nickel, B.: Z. Ges. Hyg. *30*(6), 332 (1984)
82. Chau, Y. K., Wong, P. T. S., Bengert, G. A., Kramar, O.: Anal. Chem. *51*, 186 (1979)
83. Chau, Y. K., Wong, P. T. S., Bengert, G. A., Dunn, J. L.: Anal. Chem. *56*, 271 (1984)
84. Diehl, K.-H., Rosopulo, A., Kreuzer, W.: Fresenius Z. Anal. Chem. *314*, 755 (1983)
85. Cappon, C. J., Smith, J. C.: Anal. Toxicol. *2*, 114 (1978)
86. Odanaka, Y., Tsuchiya, N., Matano, O., Goto, S.: Anal. Chem. *55*, 929 (1983)
87. Fukui, S., Hirayama, T., Nohara, M., Sakagami, Y.: Talanta *30*, 89 (1983)
88. Godden, R. G., Thomerson, D. R.: Analyst *105*, 1137 (1980)
89. Cheng, C.-N., Focht, D. D.: Appl. Environ. Microbiol. *38*, 494 (1979)
90. Johansson, B., Ryhage, R.: Acta Chem. Scand. *24*, 2359 (1970)
91. Simpson, C. F., Whittaker, M. (ed.): Electrophoretic Techniques, London—New York, Academic Press, Inc. 1983
92. Boček, P., Foret, F.: J. Chromatogr. *313*, 189 (1984)
93. Deyl, Z. (ed.): Electrophoresis, Vol 18B, Amsterdam—Oxford—New York, Elsevier Scientific Publishing Company 1983
94. Kwong, T. C.: Clin. Chim. Acta *151*, 193 (1985)
95. Berman, E.: Toxic metals and their analysis, in: Heyden International Topics in Science (ed.) Thomas, L. C., London—Philadelphia—Rheine, Heyden & Son Ltd, 1980
96. van Loon, J. C.: Analytical Atomic Absorption Spectroscopy, New York—London—Toronto—Sydney—San Francisco, Academic Press 1980
97. Ottaway, J. M.: Heavy metals determinations by atomic absorption and emission spectrometry in: Analytical Techniques for Heavy Metals in Biological Fluids, (ed.) Facchetti, S., Amsterdam—Oxford—New York Elsevier 1983
98. Delves, H. T.: Prog. Analyt. Atom. Spectrosc. *4*, 1 (1981)
99. van Loon, J. C.: Anal. Chem. *51*, 1139A (1979)
100. Sakai, T., Hanmura, S., Winefordner, J. D.: Anal. Chim. Acta *170*, 237 (1985)
101. Barnes, R. M.: CRC Crit. Rev. Anal. Chem. *7*, 203 (1978)
102. Gast, C. H., Kraak, J. C., Poppe, H., Maessen, F. J. M. J.: J. Chromatog. *185*, 549 (1979)
103. McCarthy, J. P., Caruso, J. A., Fricke, F. L.: J. Chromatogr. Sci. *21*, 389 (1983)
104. Kirkbright, G. F.: Sample introduction, signal generation and noise characteristics for argon inductively-coupled plasma optical emission spectroscopy in: Instrumentelle Multielementanalyse (ed.) Sansoni, B., Weinheim, VCH 1985
105. Uden, P. C.: J. Chromatogr. *313*, 3 (1984)
106. Carnahan, J. W., Mulligan, K. J., Caruso, J. A.: Anal. Chim. Acta *130*, 227 (1981)
107. Brainina, Kh. Z.: Stripping Voltammetry in Chemical Analysis, New York—Toronto, Halsted Press, John Wiley & Sons 1974
108. Galus, Z.: Fundamentals of Electrochemical Analysis, New York—London—Sydney—Toronto, Ellis Horwood Ltd. John Wiley & Sons 1976

ISBN 3-05-500371-3